ELEMENTARY STATISTICAL TECHNIQUES

ELEMENTARY STATISTICAL TECHNIQUES

Freeman F. Elzey

Brooks/Cole Publishing Company
Monterey, California 93940

Brooks/Cole Publishing Company
A Division of Wadsworth, Inc.

© 1985 by Wadsworth, Inc., Belmont, California 94002.
All rights reserved.
No part of this book may be reproduced, stored in a retrieval system,
or transcribed, in any form or by any means—
electronic, mechanical, photocopying, recording, or otherwise—
without the prior written permission of the publisher,
Brooks/Cole Publishing Company, Monterey, California 93940,
a division of Wadsworth, Inc.

Printed in the United States of America

10 9 8 7 6 5 4 3 2 1

Library of Congress Cataloging in Publication Data
Elzey, Freeman F.
    Elementary Statistical Techniques.
    Includes index.
    1. Statistics.   I. Title.
QA276.12.E49   1985   001.4'22   84-23826
ISBN 0-534-04668-1

Sponsoring Editor: C. Deborah Laughton
Editorial Assistant: Mary Tudor
Production Editor: Joan Marsh
Manuscript Editor: Margaret Hill
Permissions Editor: Carline Haga
Interior and Cover Design: Vicki Van Deventer
Cover Illustration: Andrew Myer
Art Coordinator: Rebecca Tait
Interior Illustration: Tim Keenan
Typesetting: Syntax International
Printing and Binding: Maple-Vail Book Manufacturing Group

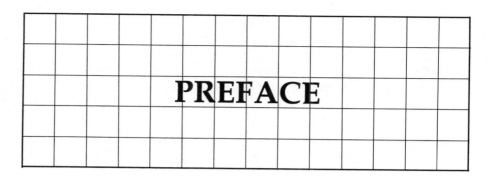

PREFACE

This book presents the fundamental statistical techniques that are generally taught in beginning courses at the undergraduate level. It represents an applied approach to the use of statistics for describing and analyzing data generated through research processes.

The focus of this book is on the major concepts and techniques covered in an introductory one-semester course; it describes commonly used statistical procedures and shows how these procedures can be used to solve problems in the behavioral sciences. Descriptive and inferential statistics are included, and fully worked-out examples are presented to illustrate each type. Methods using both numbers and graphs are employed to describe sets of data and relative placement of individual measures, formulas for estimating population parameters from sample data, and the calculation of confidence intervals. Inferential statistical methods commonly used in hypothesis-testing research involving both directional and nondirectional hypotheses are illustrated. Type I and Type II errors are covered, along with considerations of test power. The techniques discussed are the most commonly used parametric and nonparametric tests. Conventional symbolic notation has been used throughout the book.

This book is directed primarily to the student who is unfamiliar with the basic concepts of statistical techniques and the mathematics needed to apply these techniques. Only the most rudimentary algebraic skills are required, involving substitution of numbers for symbols and the solving of equations. Nothing more than addition, subtraction, multiplication, and division is needed.

The book does not attempt to develop mathematical derivations or formal mathematical proofs of the various formulas; the focus is on understanding the techniques and applying them to research problems. Each technique is illustrated by a simple example, and step-by-step calculations are shown for most techniques. Theoretical concepts are presented in narrative form. All data presented in this book are fictitious and were developed specifically to illustrate the various techniques. The data presented have been kept to the minimum amount necessary to illustrate the techniques. The student should be cautioned, however, that studies are seldom conducted in which the sample sizes are as small as those used in the examples and exercises in this text.

I am grateful to many colleagues who have used my earlier statistics texts throughout the years. Among them are Professors Hal Johnson, Enoch Sawin, Jack Frankel, Frank Hovell, and Jerry Podell, all of San Francisco State University. Their comments and suggestions are reflected in the organization and presentation of the material in this text.

I am grateful to the Literary Executor of the late Sir Ronald A. Fisher, F.R.S., to Dr. Frank Yates, F.R.S., and to Longman Group Ltd., London, for permission to reprint tables from their book *Statistical Tables for Biological, Agricultural and Medical Research (6th edition, 1974)*.

I would like to thank Richard Lohmiller, a good personal friend, who was responsible for my introduction to the world of microcomputers.

I am particularly indebted to the late Professor Samuel Levine, for his friendship and wise counsel. Without his advice and encouragement, I may never have written my first texts in statistics.

Thanks are also due to Michael Needham and Joan Marsh of Brooks/Cole Publishing Company for their consistent professional help in the development of this book. I am very grateful to Margaret Hill, whose excellent copyediting of the text did much to improve its readability.

Freeman F. Elzey

CONTENTS

1. **Introduction** 1

    Allaying Anxiety 1
    What Is Statistics? 2
    Statistics as a Tool of Research 3

2. **Variables and Levels of Measurement** 5

    Types of Statistical Measures 5
    Levels of Measurement 6
    Types of Variables 10
    Summarizing Data 11
    Exercises 15

3. **Measures of Central Tendency** 17

    The Mode 17
    The Median 18
    The Mean 20
    Exercises 24

4. **Display of Data** 27

    Pie Charts 27
    Bar Graphs 29
    Bar Histograms 29
    Frequency Polygons 31
    Types of Distribution Curves 32
    Exercises 33

5. **Populations and Samples** 35

    Population 35
    Sample 36
    Exercises 38

6. Measures of Variability   39

The Range   39
The Average Deviation   40
The Variance   41
Degrees of Freedom   44
Exercises   45

7. The Normal Distribution   49

The Normal Curve   49
The z Score   53
Exercises   56

8. Probability   57

Exercises   61

9. The Distribution of Sample Means   63

Estimating the Standard Error of the Mean   65
Exercises   67

10. Establishing Confidence Intervals   69

The 95% Confidence Interval   71
The 99% Confidence Interval   73
The t Distribution   74
Exercises   78

11. Correlation   81

Computation of the Correlation Coefficient   86
Conclusion   87
Exercises   88

12. Regression   91

The Regression of Y on X 92
The Regression of X on Y 96
Multiple Regression   99
Exercises   102

13. Introduction to Hypothesis Testing   105

Decisions Regarding the Null Hypothesis   107
Testing the Null Hypothesis   108
The Standard Error of the Difference between Means   109
Exercises   112

14. Testing for the Difference between Population Means   113

Testing the Difference between Independent Means—Variances Assumed Equal   114
Testing the Difference between Independent Sample Means—Variances Not Assumed Equal   119
Testing the Difference between Dependent Means   121
Exercises   125

15. Making Decisions about the Null Hypothesis   127

One- and Two-Tail Tests   128
Testing the Significance of a Correlation Coefficient   132
Directional versus Nondirectional Hypotheses   133
The Level of Significance   134
Type I and Type II Errors   135
The Power of Statistical Tests   136
Exercises   139

16. Analysis of Variance—One Way   141

The F Distribution   147
Tests of Multiple Comparisons   150
The Tukey Method—Equal-Sized Samples   151
The Scheffé Method—Unequal-Sized Samples   152
Exercises   154

17. Two-Way Analysis of Variance and Other Techniques   157

Two-Way Analysis of Variance   157
Analysis of Covariance   162
One-Way Analysis of Variance with Repeated Measures   166

18. Chi-Square Tests   169

The Test for Goodness of Fit   170
The Test for Independence of Two Variables   175
The Test for Equality of Proportions   177
Exercises   180

19. Nonparametric Techniques—Ordinal Data   183

Spearman Rank Correlation   184
The Mann-Whitney U Test   189
The Wilcoxon Matched-Pairs Signed-Ranks Test   191
The Kruskal-Wallis One-Way Analysis of Variance by Ranks Test   193
The Friedman Two-Way Analysis of Variance by Ranks   194
Exercises   197

Appendix

Tables 201
Answers to Exercises 219
Glossary 231
Index 233

ELEMENTARY STATISTICAL TECHNIQUES

 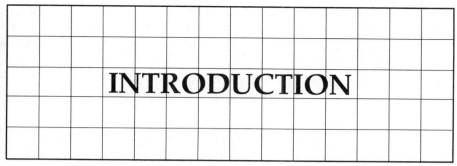

INTRODUCTION

A major obstacle that an instructor facing a group of students in a beginning statistics class must overcome is the apprehension and often downright fear of some students at the prospect of taking statistics. This apprehension has a number of causes. First, there is the fear of the unknown: these students do not have a clear idea of what statistics is all about, why they need to gain competence in it, and how this competence will be useful in their professional careers.

Second, because of the mathematical nature of the subject, students fear that it will demand more mathematical sophistication than they possess. Such fears are revealed in such expressions as "I was never any good in math," "I have forgotten all I ever knew about algebra," and "I managed to pass geometry and trigonometry but didn't learn a thing." Third, as the beginning student flips through the statistics text, he or she sees many unfamiliar terms, such as standard deviation, regression coefficient, and standard error of the mean, as well as an overwhelming array of symbols and formulas, and immediately despairs of mastering the content of the text in the short time of one semester.

Last, students in the behavioral sciences, including education, psychology, and anthropology, are sometimes biased against quantitative approaches to the study of people. They are skeptical about describing behavior in numerical terms, performing statistical manipulations on sets of numbers, and making sweeping statements about human behavior on the basis of statistical findings. If the instructor wants to teach statistics effectively, he must deal with these anxieties at the outset of the course.

ALLAYING ANXIETY

Many instructors spend part of the first class session trying to calm students who feel that they are in for an unpleasant and frustrating experience. This text is designed to help the instructor dispel these fears.

Many texts that are used in beginning statistics courses include much more material than students can assimilate during a single semester. As a result, certain sections or chapters must be skipped, breaking the author's sequence of topics. This text includes only those aspects of statistics that are generally covered in a one-semester course, and it is designed so that later topics build directly from preceding ones. Therefore, it should be used in its entirety and in the sequence presented.

The language of the text has been kept as simple and direct as possible. Some students benefit by having the more difficult statistical concepts presented in several slightly different ways. Therefore, the presentation may appear to be redundant at some points. Hopefully, students who understand the concepts on first reading will not be bothered by later repetition of them.

For most techniques discussed in this text, there is first a discussion of the rationale underlying its use, next a presentation of the formula or formulas used in it, and finally an illustration of the computations involved, using a simplified example. Many formulas are presented in two ways: the conceptual form shows the student what the formula is "doing" to the data, and the mathematically equivalent computational formula provides an easier format for the actual calculations.

All formulas are presented in the text. The appendix contains the statistical tables needed and provides a glossary of symbols and abbreviations.

The features of this text are designed to reduce the anxiety of students and to make the presentation a logical, thorough, and understandable one for beginning students. For those who need or want more guidance and practice in most of the concepts presented in this book, a supplemental self-guiding text, *A Programmed Introduction to Statistics, Second Edition,* by Freeman F. Elzey (Monterey, Calif.: Brooks/Cole Publishing Company, 1971) will prove very helpful.

WHAT IS STATISTICS?

The study of statistics provides us with methods for describing and summarizing research data, for specifying the probability that data obtained from a sample or samples come from some given or hypothesized population, for uncovering the relationship between sets of measurements, and for making predictions.

Statistics can be divided generally into two types: descriptive and inferential. *Descriptive statistics* provide us with ways to reduce quantities of data into manageable form and to describe them precisely in terms of averages, differences, relationships, and so on. The early chapters of this text present some ways of statistically describing and graphically depicting research data. However, in research in the behavioral sciences, mere description of the data obtained from a sample or samples is of little use. The scientist is more likely to be interested in the process of making generalizations from the sample to the wider population. *Inferential statistics* provide the analytic methods for making

INTRODUCTION

these generalizations. For example, if two samples of students are given two different instructional methods, and the average final scores of the two groups are different, this difference may be due either to the different instructional methods or to chance factors. Inferential statistical methods permit us to determine the probability that the difference is due to chance, rather than to the effects of the instructional methods.

> **DESCRIPTIVE STATISTICS**
> Statistical procedures used for summarizing and presenting data of samples or populations.
>
> **INFERENTIAL STATISTICS**
> Statistical procedures used for making generalizations regarding the characteristics of populations based on data from samples taken from that population.

The basic purposes of inferential statistical methods are (1) to estimate unknown population values from the observable measurements obtained from samples and (2) to test research hypotheses, using sample data. Some of the common inferential techniques used for these two purposes are presented in this text.

STATISTICS AS A TOOL OF RESEARCH

It must be stressed at the outset that statistical procedures deal with numbers. Where the numbers come from and what they represent fall in the province of the researcher. Meaningful statistical conclusions can result only from the analysis of data that have been collected through carefully controlled research studies. Where this is the case, statistics is a valuable tool in research.

Research is generally defined as a controlled inquiry used to uncover relationships among phenomena. The appropriate research design must be selected if valid conclusions are to be reached. Research projects in the behavioral sciences rely heavily on statistical procedures for data collection, organization, and analysis. In fact, assuming that the researcher has used appropriate research procedures, it is the use of statistics to present and analyze the data that provides the basis for support or nonsupport of the researcher's hypothesis. Use of proper statistical procedures is vital if the research results are to be clearly and unambiguously interpreted.

Although it is true that the primary statistical functions cannot occur until the data are collected, it would be a mistake for the researcher to ignore the talents of a statistician in planning and conducting research studies. It is of utmost importance that the statistician develop plans for organizing, summarizing, and analyzing data at the time the project is being designed. Failure to do so may lead to inappropriate and/or inefficient data collection procedures, resulting in a body of data that cannot be properly analyzed. Without statistical planning, the researcher may make costly mistakes and arrive at invalid conclusions.

In summary, this text introduces a number of statistical techniques that are generally used in research studies. The tasks of stating research hypotheses, developing research designs, selecting measuring instruments, and specifying data collection procedures are left to texts that deal with research methods.

2 VARIABLES AND LEVELS OF MEASUREMENT

TYPES OF STATISTICAL MEASURES

Statistical data are the "ingredients" with which statisticians work. A *datum* is the record of a single observation. Here the term *observation* is used in the broad sense. It may be an individual's test score, a "yes" or "no" vote for a candidate, a response to an interview question, the time it takes a rat to run a maze, or the position in which a horse finishes a race. Each observation is converted to a numerical representation since, to be useful for statistical description and analysis, the outcome of an experiment or research project is usually stated in the form of numbers. A set of numbers representing records of observations is termed *statistical data.* Thus the political preferences of a selected sample of voters, the number of runs scored by a baseball team, and the heights of sixth-grade students in San Francisco are each considered to be a set of statistical data.

The numbers constituting a set of data are quantitative representations of what we observe directly or infer from observations. These numbers can result from various types of measurement. Thus measurement techniques provide us with a process for transforming observations or inferences into usable numbers.

In the preceding examples it should be obvious that there are various kinds of measurement used for different purposes. Clearly the number of earned baseball runs, the heights of children, and the order in which horses finish a race are detemined by different measurement techniques.

Specific terms are used in research to describe the behavior or characteristic to be measured. A behavior or characteristic that takes on different values is called a *variable*. For example, if we obtain a set of spelling scores, we say that we have data on the variable of spelling ability. If we determine the sex of each member of a selected group of people, we have data on the variable of sex. Motivation, ethnic identification, social competence, age, typing ability, and so on—all are called variables. Some variables take on quantitative differences, whereas others differ in quality. Any characteristic for which the data varies among the members of the group being measured is termed a variable.

> **DATA**
> A set of numbers representing records of observations or measurements.
>
> **VARIABLE**
> A behavior or characteristic that can take on different values.

A characteristic for which the data does not vary is termed a *constant*. If a research study is concerned only with college freshmen, the class level is considered to be a constant. In most research studies we deal with both variables and constants; that is, we may select individuals for study who have certain known characteristics (constants) and then obtain data on other characteristics (variables) that we wish to learn about. Consider a study of the aggression levels of fourth-grade boys under two types of adult supervision. Here sex and grade level are constants, whereas level of aggression and adult supervision are variables. It should be evident that the same characteristics can be constants in some situations and variables in others.

What is measurement and how do we use it to transform observations into data? Measurement is the assignment of numbers to observations according to some prearranged rules. Thus, when we measure a student's arithmetic ability, we assign a number to the student that reflects the pattern of his or her responses to a given set of arithmetic problems. The manner in which each item is judged correct or incorrect and the way in which the total score is established are determined by a set of rules of measurement; that is, the items may be valued at one point each, or items may have differential weightings according to their relative importance.

LEVELS OF MEASUREMENT

The different types of measurement used to convert observations into numerical data are generally divided into four categories, which are called *levels of measurement*. Each of the following four levels is uniquely useful to the statistician.

Nominal measurement

This level of measurement involves the process of classifying objects, people, responses, and so on into categories. Thus it is used to classify people by sex, racial affiliation, or nationality, for example. In nominal measurement we list the various category headings and then determine how many observations fall under each. The categories have no logical order; their listing does not imply any hierarchical difference among them. For example, we can classify people accord-

TABLE 2-1. Political affiliations of 23 individuals

Political affiliation	Number of individuals
Republican	7
Democrat	8
Independent	3
Socialist	3
Undecided	2

ing to their political affiliation, as in Table 2-1. In the table, political affiliation is the variable being considered, and each individual in a hypothetical survey has been placed in one of the five categories.

When employing nominal measurements we must follow three rules.

1. *The list of categories must be sufficient to cover all of the observations in the study.* Every observation must fall into one of the listed categories. Thus, the categories must be exhaustive. (In our example, if we did not have a category titled "Undecided," we would have been unable to account for the responses of two individuals in our survey.)
2. *The categories must be mutually exclusive.* The descriptions of the categories must be such that any individual observation can be placed in only one category.
3. *No order is implied in the listing of the categories.* The headings only indicate different categories within the variable. The order in which they are listed is arbitrary and does not indicate any quantifiable differences among them. (In our example, we could just as meaningfully have listed them in any other order.)

For convenience, numbers are sometimes assigned to the categories, especially when the data are to be processed by computer. (In our example, the number 1 could have been assigned to Republicans, 2 to Democrats, and so on.) It is important to recognize that the number is assigned for identification only and does not indicate that one category is "better" than another. This procedure for grouping data into categories is frequently called "using a measurement scale," but this is actually a misnomer because no scale can be implied in the listing of the categories.

Additional examples using nominal measurement are given in Table 2-2.

TABLE 2-2. Frequency distributions of nominal data

(a)		(b)		(c)	
Religious affiliation	Number of individuals	Sex	Number of individuals	Automobile type	Number of individuals
Protestant	30	Female	89	Sedan	12
Catholic	45	Male	41	Convertible	3
Jewish	60			Station Wagon	5
Other	14				

> **LEVELS OF MEASUREMENT**
>
> **Nominal**
> The classification of objects, people, or observations into categories where no ordering of categories is implied. Data are numbers representing frequencies of occurrence within unordered categories.
>
> **Ordinal**
> The ordering of measurements or categories of observations and the assigning of numbers indicating the ranks. Data are numbers representing the order of individuals or measurements.
>
> **Interval**
> The measurement of phenomena by assigning numbers to observations. Data are numbers where the intervals between numbers represent equal quantities.
>
> **Ratio**
> The measurement of phenomena by assigning numbers to observations. Data are numbers where the intervals between numbers represent equal quantities *and where zero indicates total absence of the phenomena being measured.*

Ordinal Measurement

The ordinal level of measurement can be used when we are able to detect *degrees of difference* among the observations. This scale implies that there is an "ordering" of the data. In ordinal measurement, the data are arranged in rank order and they are assigned numbers representing rankings. For example, if we line up students according to weight, and assign 1 to the heaviest, 2 to the next heaviest, 3 to the next, and so on, these rankings constitute a set of ordinal measurements on the variable of weight. Our ordinal data might look like those presented in Table 2-3. Note that these numbers indicate neither how much each student weighs nor the amounts of weight differences among them. Ordinal measurement only indicates the placement of the individuals relative to each other. Other examples of ordinal measurement are the standings of football teams, order of finishing a task, and teacher rankings of students according to their academic potential.

TABLE 2-3. Rank order of weights of five students

Name	Rank
Mary	1 (heaviest)
Peter	2
Sam	3
Sally	4
Mike	5 (lightest)

Interval measurement

When we assume that the differences between our measurement units are the same size throughout the scale, we are using the interval level of measurement. In interval scales, the equal intervals or distances between units on the scale represent empirically equal differences in the characteristic being measured. This property of equal intervals permits us to perform arithmetic operations of addition and subtraction on data of this type. Not many measurements meet this stringent requirement. Although IQs are sometimes treated as though they were on an interval scale, equal units of IQs do not actually represent equal increments in intelligence. For example, the difference between IQs of 130 and 140 represents a much larger increase in intelligence than does the difference between IQs of 100 and 110.

A distinguishing characteristic of this scale is the fact that the zero point does not necessarily represent the total absence of the phenomenon being measured. For example, a score of zero on a statistics quiz does not mean that a person is devoid of all knowledge of statistics, and a score of zero on a college aptitude examination does not mean that the person taking the test has no aptitude for college at all! Nor is it logical to say that a person having an IQ of 120 is twice as intelligent as a person having an IQ of 60.

Generally the test developer arbitrarily selects the number that is assigned to a given level of performance. (The number 400 could have been designated the average IQ just as easily as 100 was.)

A commonly cited example of interval measurement is the Fahrenheit scale for measuring temperature, in which zero does not represent the total absence of heat. It would be inaccurate to say that on this scale 100° is twice as hot as 50°.

Ratio measurement

In contrast to the interval scale, the ratio scale does assume that there is an absolute zero point at which the scale originates. In cases where zero represents the total absence of a phenomenon and equal-sized units from this zero point represent empirically equal differences, we are using the ratio scale of measurement. Common ratio scales are those measuring weight, time, and height. On such scales it is possible to indicate that one person weighs twice as much as another or that it took one rat four times longer than its companion to run a maze. Ratios between such measurements can be meaningfully interpreted. For example, on the Kelvin scale for measuring temperature, zero indicates the total absence of heat and a temperature of 100° is considered to be twice as hot as one of 50°.

Note that very few variables in educational and psychological studies lend themselves to the ratio scale of measurement. Also, many measurements, such as test scores, are commonly treated as interval-scale measurements even though the assumption of equal-sized intervals may be questionable.

It is important that the statistician determine whether the data have been obtained by counting (nominal measurement), by ranking (ordinal measurement), or by measuring quantities (interval or ratio measurement), because different statistical approaches are used with the different types of measurement scales. In this text we present statistical techniques that are appropriate for obtaining data using each of these scales. For our purposes, we refer to measurements on the interval and ratio scales as simply *interval data* because they are treated the same in our applications.

TYPES OF VARIABLES

At this point we need to consider one other characteristic of statistical data that affects the way in which we analyze them. There are two types of variables—*discrete* and *continuous*. Discrete variables are those variables that can take on only a finite number of values between any two points. A variable such as the size of families is a discrete variable because we do not consider a family as having, say, $3\frac{1}{4}$ or $6\frac{1}{2}$ members. This variable has only four possible values between 3 and 6. Similarly, the number of houses on city blocks is considered a discrete variable because only a finite number of values can lie between any two given points on the variable. For example, between 10 and 20, there can only be 11 possible values for "number of houses."

TYPES OF VARIABLES
Discrete
A variable that can assume only a finite number of values between any two points.
Continuous
A variable that theoretically can assume an infinite number of values between any two points.

Variables such as weight, length, and age, for which measurements can theoretically have *any* value, are considered *continuous* variables. For example, a person's age or weight can be represented by any one of an infinite range of values; the actual value we assign is determined by the precision of our measuring instrument. Thus a man's weight may be 165.4124 pounds (or any other fractional value between 165 and 166 pounds), and he may be 29.6803 years old (or any other value between 29 and 30 years). Even these measurements are probably inexact; the fractional portion could be extended even further with more sensitive measuring devices. We are never able to obtain exact measurements on continuous variables; in contrast to measurements on discrete variables, which can be exact, measurements on continuous variables can only be approximate.

VARIABLES AND LEVELS OF MEASUREMENT

Psychological variables, such as intelligence, motivation, anxiety, and desire for approval, are theoretically considered continuous, even though we must settle for discrete approximations in our measurements of them. For example, if Mary obtains an IQ of 112, we must consider this an approximation of her intelligence quotient, which, barring measurement errors, we assume to be somewhere between 111.5 and 112.5.

In summary, when we use discrete variables, we are dealing with exact measurements; when we use continuous variables, we are always obtaining approximate measurements. We use different statistical techniques depending on whether the data represent continuous or discrete variables.

SUMMARIZING DATA

Our first task when we are confronted with data is that of summarizing and organizing them into a form suitable for display and analysis. For example, suppose we ask 40 elementary schoolchildren which animal they would choose as a pet and receive the following responses:

Dog	Rabbit	Cat	Turtle	Dog
Cat	Dog	Dog	Dog	Cat
Turtle	Dog	Cat	Rabbit	Cat
Rabbit	Cat	Dog	Rabbit	Dog
Cat	Cat	Cat	Cat	Turtle
Rabbit	Dog	Cat	Dog	Rabbit
Cat	Turtle	Rabbit	Dog	Dog
Rabbit	Dog	Dog	Rabbit	Dog

It is difficult to determine the children's preference pattern from this set of data. We need to organize the data to get a clear picture of the trend of the children's responses. The first step is to determine the frequency with which each animal has been chosen. In this example, "dog" was selected by 15 children. In statistical terms we say that the category "dog" has a frequency of 15. We can determine the frequency of choice for each of the other animals and list them in a *frequency distribution*, as shown in Table 2-4. Here we have introduced our first two statistical symbols: f, meaning frequency, and N, meaning the total

TABLE 2-4. Frequency distribution of pet choices of 40 children

Animal	f
Rabbit	9
Cat	12
Turtle	4
Dog	15
	$N = 40$

number of responses. In this example, the data collected represent nominal measurements, because there is no hierarchical "scale" implied in the arrangement of the animal choices presented to the children. In preparing the frequency distribution given in Table 2-4, we could just as meaningfully list the four animals in any order.

FREQUENCY DISTRIBUTION
A table that depicts how individuals or measures are distributed within a set of categories or values.
Symbols
N Total number of individuals or scores.
f Frequency of individuals or scores.
X Value of a score.

In Table 2-4, a child chose only one from a limited number of alternatives. The children's choices represent measurements on a *discrete* variable; that is, we can consider this an example of exact measurement, because there is no possibility of a choice falling between any two alternatives.

Having collected these data, we can now *rank* the animals according to the children's preferences, thus converting our original nominal measurements to ordinal measurements. By assigning a rank of 1 to the most preferred animal, we can obtain the ordinal measurements given in Table 2-5. Note that we could not have ranked the animals before collecting the data. The distinction here is that the list of animals represents the nominal level of measurement, whereas the order of the children's preferences for the animals represents the ordinal level.

TABLE 2-5. Rank order of pet preferences of 40 children

Animal	Rank
Dog	1
Cat	2
Rabbit	3
Turtle	4

Consider another set of data. Thirty-five college students were given a test designed to assess their anxiety at the prospect of taking a course in advanced calculus. Their anxiety test scores are presented in Table 2-6.

To get an overall picture of how the students scored on this test, we can prepare a frequency distribution of the scores. In statistics a score value is generally represented by the symbol X. The frequency with which a score value occurs in a set of data is listed in the f column of the frequency distribution.

VARIABLES AND LEVELS OF MEASUREMENT

TABLE 2-6. Anxiety test scores for 35 students

71	68	69	68	70
69	69	70	68	69
70	68	72	69	70
66	71	68	70	68
69	68	69	66	69
69	69	70	70	68
70	70	69	71	69

The frequency distribution of the anxiety test scores is presented in Table 2-7. (In preparing a frequency distribution, it is customary to place the lowest value at the bottom of the distribution.)

TABLE 2-7. Frequency distribution of anxiety test scores for 35 students

X	f
72	1
71	3
70	9
69	12
68	8
67	0
66	2
	$N = 35$

In statistics X generally represents a *known* value, whereas in algebra the symbol X is commonly considered an unknown quantity. In Table 2-7, each individual has a known score that can be called his or her X score.

Scores on psychological tests such as this one usually represent measurements that fall between the ordinal and interval scales. Furthermore, the variable of anxiety is theoretically considered a continuous variable; thus it is conceivable that an individual could score at any point along the continuum. Therefore, the fact that all the reported scores on this test are integers merely reflects the inability of our test to detect small differences in levels of anxiety.

The scores we obtain on a continuous variable such as anxiety are the results of a "rounding off" process. The *real* limits of each score value lie .5 above and .5 below the integer value; thus each score value represents an interval within which the "true" score lies. For example, the *real* score of a student obtaining an anxiety test score of 69 lies somewhere between 68.5 and 69.5. To describe the distribution in Table 2-7 precisely, we can say that 12 students have anxiety test scores falling in the interval between 68.5 and 69.5. In dealing with continuous variables, we will find the lower limit of a score value useful in some future calculations. This lower limit is represented by the symbol $\ell\ell$. For example, the $\ell\ell$ for the score value 71 is 70.5; for the score value 70, the $\ell\ell$ is 69.5.

When we prepare a frequency distribution of interval measurements, we list each value from the highest to the lowest in the X column whether or not it was obtained by any individual. (In Table 2-7 the score value of 67 is listed although no one received that particular value.) When data are not grouped, a frequency distribution presents the f for each separate score.

It is sometimes desirable to group data with a wide range of scores into categories called *class intervals*, particularly when we wish to present data in tabular form or to display them graphically. Table 2-8 presents a frequency distribution of scores in which two score values have been grouped to form each class interval.

TABLE 2-8. Grouped frequency distribution of scores

Interval	f
20–21	1
18–19	0
16–17	8
14–15	8
12–13	5
10–11	2
8–9	1
	$N = 25$

When grouping the scores into class intervals, we "lose" some of our definitive information. We see from Table 2-8 that five individuals received scores of 12 or 13, but we do not know whether all five individuals received a score of 12, all received a score of 13, or there was a mixture of the two scores. The larger the class intervals we form when we prepare a frequency distribution, the greater the loss of definitiveness.

If the data in Table 2-8 represent measurement on a continuous variable, then each class interval has a real upper and lower limit, just as the individual score values discussed earlier had. The lower limit of a class interval is .5 below the lowest score value in the interval, and the upper limit is .5 above the highest score value in the interval. Thus the interval 10–11 actually ranges from 9.5 to 11.5. Its $\ell\ell$ is 9.5. In Table 2-8, the real lower and upper limits of the entire distribution are 7.5 and 21.5.

This consideration of upper and lower limits for continuous variables also pertains to distributions where fractions are used in forming class intervals. If we have a class interval of 102.59–102.60, the real lower limit of the interval would be 102.585 and its real upper limit would be 102.605.

Remember that this concern with real upper and lower limits applies only to data that have been obtained from theoretically continuous variables. If the values in Figure 2-8 had represented the number of days students were tardy for school (a discrete variable, since students are considered either tardy or not tardy), we would not need to be concerned with upper and lower limits of the obtained data.

Although there are few hard and fast procedures for preparing frequency distributions, one practice that is generally followed is to make all of the intervals the same size.

Frequency distributions that must summarize a large number of data usually have anywhere from 12 to 20 class intervals. In Table 2-7, where N was only 35, we formed only seven class intervals. As said earlier, the grouping of data into class intervals is primarily useful for presenting them in frequency distributions or in graphic form.

EXERCISES

1. A group of children gave the following responses when asked to name their favorite ice cream flavor. Prepare a frequency distribution of these data.

Chocolate	Vanilla	Chocolate	Vanilla	Chocolate
Vanilla	Chocolate	Vanilla	Chocolate	Chocolate
Vanilla	Vanilla	Vanilla	Chocolate	Strawberry
Strawberry	Vanilla	Rocky road	Rocky road	Strawberry
Rocky road	Strawberry	Strawberry	Vanilla	Rocky road

2. What level of measurement is represented by the data in Exercise 1?
3. Using the data in Exercise 1, what is the rank order of the children's preferences for ice cream flavors?
4. What level of measurement is represented by the ranks in Exercise 3?
5. What do the following symbols represent? $X, f, N, \ell\ell$
6. Prepare an ungrouped frequency distribution of the following spelling test scores.

18	24	19	15
13	27	15	23
19	17	22	19
22	20	28	16
21	14	21	20
9	26	14	

7. Prepare a grouped frequency distribution of the spelling achievement test scores in Exercise 6, using an interval size of 3 points.
8. What level of measurement is represented by the data in Exercise 7?
9. Prepare a frequency distribution for the following set of raw scores. Let the interval size be 4 points. What is the N of this distribution?

31	33	23
29	44	30
35	33	36
40	21	34
38	37	35
38	41	

10. What is the $\ell\ell$ of each of the following? 12, 20, 17–19, 100–104

3 MEASURES OF CENTRAL TENDENCY

Data that have been collected and organized into a frequency distribution are ready for further statistical treatment. Our first concern in describing a set of statistical data may be to determine the one score value that best characterizes the entire frequency distribution. When there are many scores in a frequency distribution, they generally tend to group at some central or representative value; this value is called a *measure of central tendency*. We will examine three measures of central tendency—the mode, the median, and the mean.

THE MODE

The mode is defined as the most frequently occurring score value in a frequency distribution; it is the value that has the greatest frequency. In column (a) in Table 3-1 the value 38 is the mode of the frequency distribution because that score was obtained by more individuals than was any other score. The mode of the distribution in column (b) of Table 3-1 is the value 99, because that score was received by the most individuals. A frequency distribution may have more than

TABLE 3-1. Three frequency distributions of ungrouped data

(a)		(b)		(c)	
X	f	X	f	X	f
41	1	100	7	60	9
40	4	99	10	59	12
39	5	98	9	58	10
38	9	97	8	57	11
37	2	96	6	56	8
36	1	95	6	55	12
		94	2	54	9
		93	0	53	6
		92	1	52	2

one mode; the distribution in column (c) of Table 3-1 is called *bimodal* because it has two values, 59 and 55, that have highest frequencies of occurrence. A distribution that has many modes is called *multimodal*.

The mode is not necessarily a value near the center of a frequency distribution, as column (b) in Table 3-1 illustrates. It can be a high or a low score value. The mode of a distribution tells us nothing about the range of scores or their variability in the distribution. It only tells us which score value or values occurred most frequently.

In a frequency distribution of grouped data, the interval that occurs most frequently is called the *modal interval* of the distribution. In Table 3-2 the interval 61–63 is the modal interval.

TABLE 3-2. Frequency distribution of grouped data

Interval	f
70–72	7
67–69	12
64–66	12
61–63	14
58–60	11
55–57	9
52–54	8
49–51	4

Since the mode need not be centrally located in the distribution, it is not a very useful measure of central tendency but is a good measure of the typical case. The mode also tends to vary more from one sample to another than do the other measures of central tendency that will be discussed.

THE MEDIAN

The median is the value in a frequency distribution that divides the *frequency* of scores in half; that is, 50% of the scores are above the median and 50% are below it. The median of each of the frequency distributions in Table 3-3 is 13, because 5 people in each distribution have scores above 13 and 5 people have scores below 13. Obviously these three distributions are quite different. The scores in (a) are scattered, whereas in (c) they are very close together. These distributions illustrate the point that, although the median is a measure of central tendency, to determine it we need not take into account the values of the scores above or below it. It is determined only by how many scores are on either side of it.

It is easy to determine the median of the distributions in Table 3-3 by inspection. But how can we determine the median of the distribution in Table 3-4? For most purposes, it is accurate enough to simply report the score value that contains the midpoint of the distribution as the median. In Table 3-4 this

TABLE 3-3. Three frequency distributions

(a)		(b)		(c)	
X	f	X	f	X	f
17	1				
16	0	16	1		
15	3	15	1	15	3
14	1	14	3	14	2
13	1	13	1	13	1
12	2	12	2	12	5
11	1	11	2		
10	0	10	1		
9	1				
8	1				

TABLE 3-4. Frequency distribution

X	f
72	1
71	3
70	9
69	12
68	8
67	0
66	2
	$N = 35$

value is 69. Twelve individuals received that score and the median person must be among them. If we assume that this frequency distribution represents measurements on a continuous variable, then the value of 69 must have a lower limit of 68.5 and an upper limit of 69.5. There is a formula for locating the point between 68.5 and 69.5 that precisely marks the median, but we need not be this precise.

In Table 3-4, because there are 35 individuals, the position of the median individual is determined by taking one-half of 35, or 17.5. Thus we count the frequencies from the bottom of the distribution until we find the value that contains the 17.5th individual. In this example, the median is 69 because the middle two individuals (17th and 18th) in this distribution both received a score of 69. In Table 3-5, the median is 46.5 because this point divides the distribution into 11 individuals in each half.

As we will see later, the median represents the 50th percentile. The median generally gives us more information about the nature of the distribution of scores than does the mode, because it gives us the score that divides the frequencies into two halves, whereas the mode tells us nothing about the other scores in the distribution. For example, if you are told that the modal score on a midterm examination is 97, you know nothing about the distribution of scores.

TABLE 3-5. Frequency distribution

X	f
49	1
48	3
47	7
46	6
45	3
44	2
	$N = 22$

If you are told that the median score is 75, you know that 50% of the students scored above that score and that 50% scored below it, but you still do not know how the scores were "spread" around 75.

THE MEAN

A measure of central tendency that takes into account the *value* of each score is called the *arithmetic mean*. The arithmetic mean of a distribution is what is commonly called the *arithmetic average* of all the scores. We refer to it simply as the *mean*. It is obtained as you ordinarily obtain an average; that is, by adding all the scores together and dividing this total by the number of scores in the distribution. Formula 1 is the formula for the calculation of the mean.

FORMULA 1

Calculation of the mean.

$$\bar{X} = \frac{\Sigma X}{N}$$

Formula 1 introduces the symbol for the sample mean, $\bar{X}$ (read as "X bar"). Formula 1 also introduces another new symbol, the capital Greek letter Σ, which is pronounced "sigma" and means "the sum of." The symbol ΣX indicates that we sum all of the X scores. This sum is then divided by N to obtain the mean.

For the distribution in Table 3-6, the frequency of each X is given in the second column (f). The bottom of the third column gives $\Sigma X = 2208$, determined by summing the values of each X times its frequency. For example, 4 individuals had a score of 75, which contributes 300 points (4 times 75) to the total sum. To determine the mean of this distribution, we substitute the values for the symbols in Formula 1 and solve for $\bar{X}$.

$$\bar{X} = \frac{2208}{30} = 73.6$$

MEASURES OF CENTRAL TENDENCY

TABLE 3-6. Frequency distribution

X	f	
78	1	78
77	2	154
76	0	0
75	4	300
74	7	518
73	9	657
72	4	288
71	3	213
	N = 30	ΣX = 2208

The mean of the distribution in Table 3-6 is 73.6. What does the mean of a distribution really represent? The mean is the unique point in a distribution that represents the "balance point" of the data. If we think of the data in Table 3-6 as weights on a calibrated teeter-totter, we might picture them distributed as in Figure 3-1.

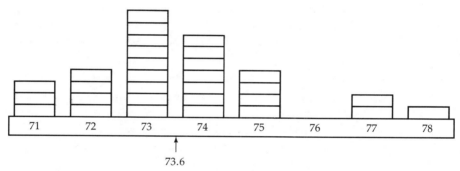

FIGURE 3-1. Graphic display of frequency distribution in Table 3-6.

The point at which the teeter-totter would balance is 73.6. This figure illustrates an important property of the mean: the sum of the deviations of the smaller scores to the left of the mean is equal to the sum of the deviations of the scores to the right of the mean. For the distribution in Table 3-6, we can compute the sum of the deviations from $\bar{X}$ as shown in Tables 3-7 and 3-8.

TABLE 3-7. Sum of deviations for $X > \bar{X}$

X values larger than $\bar{X}$	$\bar{X}$	Deviation of X from $\bar{X}$	f	Deviation values
78	73.6	4.4	1	4.4
77	73.6	3.4	2	6.8
76	73.6	2.4	0	0
75	73.6	1.4	4	5.6
74	73.6	.4	7	2.8
			Sum of deviations =	19.6

TABLE 3-8. Sum of deviations for $X < \bar{X}$

X values smaller than $\bar{X}$	$\bar{X}$	Deviation of X from $\bar{X}$	f	Deviation values
73	73.6	−0.6	9	−5.4
72	73.6	−1.6	4	−6.4
71	73.6	−2.6	3	−7.8
			Sum of deviations =	−19.6

Here we have shown that the sum of the deviations for scores larger than the mean is the same as the sum of the deviations for scores smaller than the mean. If we add all of the deviations we get $-19.6 + 19.6 = 0$. This is probably the most important property of the mean: that the sum of all the deviations from it is zero.

In contrast to the mode and the median, the mean is sensitive to the value of each score in the distribution. For example, if the individual in Table 3-6 who received a score of 78 had received a score of 90, neither the mode nor the median of the distribution would have been affected, but the mean certainly would have been higher!

In summary, we have learned how to determine three measures of central tendency:

1. *Mode.* The most frequently occurring score in a frequency distribution.
2. *Median.* The point in a frequency distribution above which half of the scores lie and below which half of them lie.
3. *Mean.* The arithmetic average of the scores in a frequency distribution.

Although the mean is computed by using the *value* of each score in the distribution, this is true of neither the mode nor the median: the mode is based solely on the frequency of the scores and the median is based on the relative positions of the scores, without regard to their values. Ordinarily the mean is the most useful and stable measure of central tendency, and for most sets of data it is the most representative score. Furthermore, as we will see, the mean lends itself to a wide variety of statistical manipulations; the other two measures do not.

As Figure 3-2(a) shows, the mode, median, and mean all coincide in a symmetrical distribution. This is not the case with skewed curves. In positively skewed curves, such as Figure 3-2(b), the median falls to the right of the mode and the mean lies even farther to the right. The order of these three measures of central tendency is reversed in negatively skewed curves, as is shown in Figure 3-2(c). These differences show that the value of the mean is more affected by the shape of the distribution than is the value of either the mode or the median.

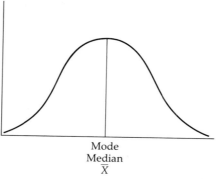
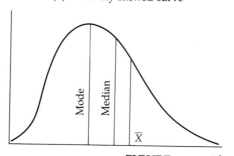
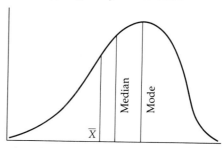

FIGURE 3-2. Three distribution curves.

As an example, suppose that a supervisor obtained production rates on a group of employees as shown in Table 3-9.

TABLE 3-9. Production rates

68	61	62	67
67	66	67	64
65	70	66	67
64	68	65	64
62	66	65	65
66	63	64	63
65	64	63	66
64	64	63	65
63	65	66	64
61	62	64	62

The frequency distribution for the data in Table 3-9 is presented in Table 3-10.

From Table 3-10 we can easily see that the mode of this distribution is 64. However, this measure of central tendency does not tell us anything about how the rates are distributed. In fact, since it is the least informative of the measures of central tendency, we will not concern ourselves with it further.

TABLE 3-10. Frequency distribution of production rates

X	f	X	f
70	1	65	7
69	0	64	9
68	2	63	5
67	4	62	4
66	6	61	2
			$N = 40$

EXERCISES

1. Twenty-two Boy Scouts were tested on their ability to read a topographical map. The following map reading scores were obtained. Prepare an ungrouped frequency distribution of these data. Determine the mode, median, and mean of these scores.

27	29	25	28
23	28	27	23
30	23	28	23
28	26	28	27
29	25	27	
26	27	28	

2. The visual acuity of the employees in the Able Manufacturing Company was tested and the following visual acuity scores were obtained. Prepare an ungrouped frequency distribution of these data. Determine the mode, median, and mean of the scores.

62	60	61	62
59	58	60	62
58	60	61	59
61	63	62	61
60	57	61	60
58	63	61	59

3. The following data show the number of trials a group of college freshmen took to recall a list of nonsense syllables accurately. Prepare an ungrouped frequency distribution of these data. Determine the mode, median, and mean for these data.

5	3	6	5
4	4	4	1
3	5	2	5
5	6	5	4
2	3	5	3
6	4	4	5

4. A group of high school graduates were given a scholastic aptitude test and the following scores were obtained. Determine the mode, median, and mean for these data. Prepare a grouped frequency distribution with each interval having a width of 5 points and the midpoint of the lowest interval set at 102.

119	115	112	105	118	110	114	122
115	107	122	123	113	118	104	100
117	128	116	112	119	120	126	116
114	119	129	105	119	110	114	101
118	117	119	124	113	117	119	105
125	104	107	119	121	116	108	126
109	124	115	123	120	113	128	115
118	106	111	117	117	113	115	108
109	115	118	118	109	100	104	
122	111	103	116	110	119	113	

4 DISPLAY OF DATA

When a distribution of scores is drawn up in tabular form, it is difficult to see the "shape" of the distribution. It is often easier to see how the scores are spread within a distribution when they are presented in graphic form.

Many types of graphs can be prepared to display statistical data. The particular type that is appropriate for a given set of data depends on the level of measurement involved and on whether the data represent discrete or continuous variables.

GRAPHIC DISPLAY MODES—NOMINAL DATA

Pie Chart
A segmented circle with a segment's area corresponding to the percentage of total frequency in a category.

Bar Graph
A series of unconnected bars, the height of each bar representing the frequency in a category.

GRAPHIC DISPLAY MODES—INTERVAL DATA

Bar Histogram
A sequential series of connected bars, the height of each bar representing the frequency of a particular value.

Frequency Polygon
A sequential series of connected points, the distance of each point from the baseline representing the frequency of a particular value.

PIE CHARTS

For nominal data, a simple frequency distribution presented in tabular form generally gives a clear "picture" of how the data are distributed. For example, the

TABLE 4-1. Frequency distribution of nominal data

Religious affiliation	f	%
Protestant	211	38.3
Catholic	194	35.3
Jewish	89	16.2
Other	56	10.2
	N = 550	

distribution in Table 4-1 presents the data in a way that can be easily interpreted. If a graph is desired, a common technique for depicting nominal data is the preparation of a *pie chart*, in which the percentage of N representing each category's frequency is used to determine the size of its corresponding segment. Figure 4-1 shows the pie chart for the data presented in Table 4-1.

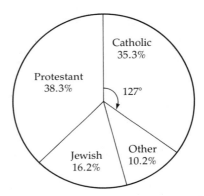

FIGURE 4-1. Pie chart of data in Table 4-1.

The area of each segment in the pie chart represents the percentage of frequencies for the corresponding category in the frequency distribution. To determine how large to make each segment, we determine the percentage of the circle (360°) that corresponds to the percentage of each frequency. In our example, 35.3% of the people in the sample are Catholic. Therefore we take 35.3% of 360°, which is 127.08°. Using a protractor, we draw a pie segment with an angle of 127.08° to represent the portion of the sample that is Catholic. The angles of the other segments of the circle are determined similarly.

In Figure 4-1, the segment for Catholics extends from 0° clockwise to 127.08°. Of course, this segment, as well as any of the others, could be placed anywhere in the circle. In a pie chart, it is customary to label each segment and to show the percentage represented by each segment rather than the degrees of their angles.

BAR GRAPHS

Another appropriate method for graphically depicting nominal measurements such as colors, ethnic group affiliations, and geographical areas is the *bar graph*. The data presented in Table 4-1 are depicted in bar graph form in Figure 4-2. Note that the heights of the bars correspond to the frequencies of the various categories, and that there are spaces between the bars in this graph. This spacing of the bars indicates that there is no continuum underlying the nominal categories. In a bar graph, the spaces between the bars should be about one-half the widths of the bars. The bars are generally arranged in the order of their magnitude, with the largest at the extreme left side of the graph and the smallest at the extreme right.

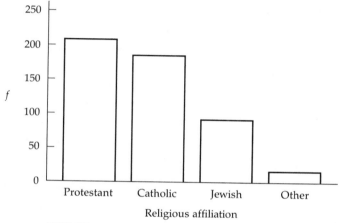

FIGURE 4-2. Bar graph of data in Table 4-1.

BAR HISTOGRAMS

A special type of bar graph, called a *bar histogram*, is used when measurements are to be depicted on the interval scale. A bar histogram resembles a bar graph, except that in it the bars are adjacent to each other. This indicates that the data depicted form a continuum. Figure 4-3 presents a frequency distribution for ungrouped discrete measurements of absences per student and its bar histogram. Note that the horizontal axis in the bar histogram in Figure 4-3 gives the discrete values for these data, and the height of each bar represents the frequency with which each discrete value occurs.

In a bar histogram depicting data from a continuous variable, the boundaries of the bars are located on the horizontal axis at the real lower and upper limits of the intervals. The grouped frequency distribution and the bar histogram showing the intelligence quotients for 110 students are given in Figure 4-4. Note that in this bar histogram, the boundaries of each bar represent the real limits of the corresponding intervals in the frequency distribution. The values indicated

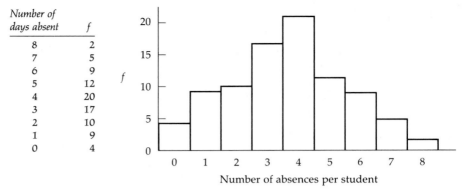

FIGURE 4-3. Frequency distribution and bar histogram of absentee rate.

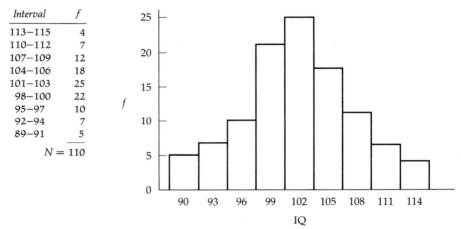

FIGURE 4-4. Frequency distribution and bar histogram of intelligence quotients.

along the horizontal axis are the midpoints of the intervals. In Figure 4-4, we see how much easier it is to get an idea of the shape of the distribution from the bar histogram than it is from the frequency distribution.

The following are conventions that are generally followed in the preparation of histograms:

1. The frequencies of the observations should appear along the vertical axis, with the zero point at the bottom of the axis.
2. Values of the variable or midpoints of intervals being depicted should appear along the horizontal axis, with the lowest value at the left of the axis.
3. Both axes must be uniformly calibrated. Even where an interval has zero frequency, space must be left in the histogram to represent zero frequency.
4. The widths of the bars must all be the same.

FREQUENCY POLYGONS

The bar histogram is used to display graphically a set of interval or ratio data. Another type of graph, called the *frequency polygon,* can be used instead to emphasize further the continuous nature of the measurement scale. In preparing a frequency polygon, we can use the same calibrations on the vertical and horizontal axes as we used for the bar histogram. Instead of representing each interval by a bar, however, we can place a dot over the midpoint of the interval and then connect all of the dots to form the polygon. The heights of the dots indicate the frequencies of the intervals. Figure 4-5 presents the frequency polygon for the data given in Figure 4-4.

In Figure 4-5, the midpoint of each interval in the frequency distribution is represented by a dot at the height that corresponds to that interval's frequency.

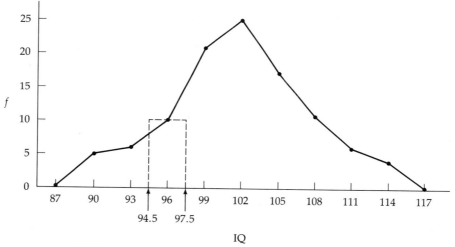

FIGURE 4-5. Frequency polygon for data in Figure 4-4.

As an illustration, the bar for interval 95–97 in the histogram is given in dotted outline. The midpoint of the interval is shown as a dot at the center of the top of the bar, directly over the IQ value of 96, which is the midpoint of the interval. In similar fashion, each dot on the frequency polygon corresponds to the center of the top of a bar from the histogram.

Usually the ends of a frequency polygon are "tied down" to the horizontal axis by extending the distribution to include intervals at each end that have zero frequencies. In the frequency polygon in Figure 4-5, the frequency of zero is shown for midpoints 87 and 117.

As shown in Figure 4-6, when many score values and many frequencies are depicted in a frequency polygon, the trend of the data, or the shape of the distribution, becomes quite apparent. In these cases, a smooth curve can be drawn

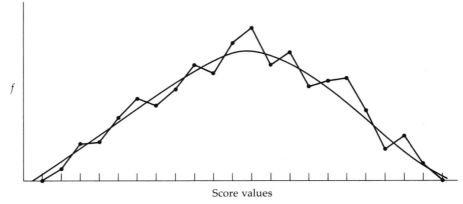

FIGURE 4-6. Frequency polygon and smoothed curve.

through the frequency polygon. Such a curve is superimposed on the frequency polygon in Figure 4-6. This curve, sometimes called a *smoothed polygon*, represents the best-fitting curved line that can be drawn through the many points, as opposed to a straight line connecting each adjacent pair of dots. This curve is an approximation of the frequency polygon; it has smoothed out the jagged edges, leaving only the trend. However, much of the jaggedness in a frequency polygon may be due to random error in the data, so that the smoothed curve may well be a more accurate depiction of the distribution of the "true" measurements than the frequency polygon is.

TYPES OF DISTRIBUTION CURVES

The shape of a distribution curve depends on the way in which the data are distributed. Figure 4-7 gives four typical curves.

The negatively skewed curve in Figure 4-7(a) represents a frequency distribution in which the frequency of large values is much greater than the frequency of small values. This curve is skewed to the left because the slope of the curve trails off toward the left end of the graph, where the lower values lie.

Figure 4-7(b) is called a positively skewed curve because the slope of this curve trails off to the right, where the higher values lie. A test that is very difficult would probably result in a positively skewed curve because few students would obtain high scores. In frequency distributions with two modes, the smoothed curve, which might look like Figure 4-7(c), is called a *bimodal curve*.

A symmetrical distribution curve, as shown in Figure 4-7(d), has some distinctive properties that are of interest to statisticians. Since the curve is symmetrical, its left side is a mirror image of its right side and the mode of the curve is at its center, dividing the distribution exactly in half. Thus half of the frequencies lie below it and half of them lie above it.

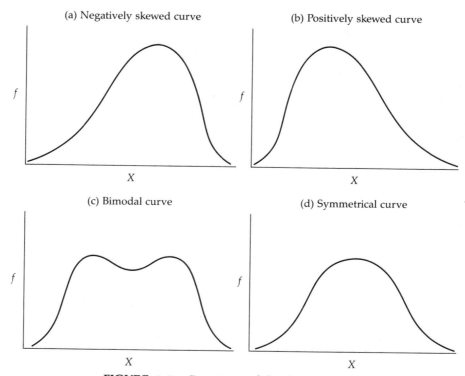

FIGURE 4-7. Four types of distribution curves.

EXERCISES

1. A survey researcher asked 182 individuals what type of transportation they used to get to their place of employment. She obtained the following data. Prepare an appropriate graphic display of these data.

Type	f
Private auto	80
Bus	35
Train	21
Car pool	46

2. Using the data in Exercise 3, Chapter 3, prepare an appropriate graphic display of the number of trials the group of college freshmen took to list the nonsense syllables accurately.

3. Using the data in Exercise 4, Chapter 3, prepare an appropriate graphic display of the scholastic aptitude scores for the high school graduates.

5 POPULATIONS AND SAMPLES

One of the important concerns of the statistician is the study of the relationship between a population and samples drawn from it.

To determine the mean height of all students currently enrolled in physical education classes at Terrence University, we might be able to arrange to measure all of them on the same day. We could then compute the mean height of this group. In this example, we would have measured the *population* defined as "Terrence University students currently enrolled in physical education classes." Having measured every student, we would know the value of the population mean height, and we would not have to make any inferences about its value. On the other hand, if it were not feasible for us to measure every student fitting the description, we would select a *sample* of the population, measure them, and compute the mean height of the sample. (Methods for selecting samples are discussed later.) We could then infer that the mean height of the sample was "very similar" to the mean height of the population. We would know that the obtained sample mean was probably not exactly equal to the population mean, because samples are seldom perfect representations of their populations: there is always some *sampling error* involved when a sample is selected from a population. This process of evaluating sampling error is one of the principal functions of statistics.

POPULATION

Usually we think of a population as a large number of people. The term *population* has a specific meaning to statisticians. Examples of statistical populations are the weights of all 6-year-old males in the U.S., the daily calorie intake rates of all 8-day-old rats, the number of rooms in all single-family dwellings in Chicago, and the arithmetic achievement scores of all sixth-grade students in the Halifax school district. The key word in the above examples is *all*. It is important for research purposes that the terms defining a population be clear and specific so that there is no confusion regarding its composition.

> **POPULATION**
> The measurements on a given variable for all members of a defined group.
>
> **SAMPLE**
> A selected subset of a population.
>
> **RANDOM SAMPLE**
> A sample in which every member in a given population has an equal chance of being selected for the sample.

Technically a statistical population does not include the individuals or objects in a defined group, but rather their measurements on a given variable. Thus, in the first example given under this heading, the statistical population is not all 6-year-old males in the U.S., but rather their weights.

Using this definition, we are not limited to describing populations in terms of numbers of people, animals, or objects. We can speak of a population of reaction times for Tom Smith (here only one person is involved) or populations of arithmetic test scores and manual dexterity ratings of third-grade pupils (thus the same group can yield two statistical populations).

SAMPLE

A sample is any selected subset of a population. Samples vary in size from one up to one less than the size of the population. The basic function of statistical inference is to analyze data drawn from a sample or samples in order to make a generalization about the population from which the samples were drawn. Therefore the researcher must select samples that are likely to be representative of the population. Of course, there is no way to ensure selection of a perfectly representative sample, but statistical analysis can provide an estimate of error in sampling.

The most important assumption underlying statistical inference procedures is that the sample or samples have been randomly selected from the population. A *random sample* is defined as one in which the selection of one member is independent of the selection of other members. Thus every member has an equal chance of being selected from a given population.

For studies in the behavioral sciences, this process is not as simple as it sounds. First, it assumes that we can identify every sample of a population and afford each sample an equal opportunity for selection. When we are dealing with very large populations, it is difficult or impossible to meet this requirement. (Consider the task of specifying each member of the population of kindergarten children in American schools.) Second, assuming that we can identify the members of a specified population, we must have a method that ensures a random selection of its members to serve as a representative sample. One way is to place each member's name or code number in a barrel, stir them thoroughly, and blindly draw out a sample. Another method is to assign each member of the population a code number, and select code numbers for the sample by using a table of random numbers in which a large collection of numbers have been prerandomized.

In most cases, neither of the preceding methods can be used to select samples for studies in the behavioral sciences. Usually we must settle for some pre-existing group to serve as a sample—such as when we study Mr. Johnson's first-grade class and assume that it is representative of all first-grade pupils. This may or may not be a warranted assumption. On the other hand, in studies where two or more samples are selected to compare the effects of treatment, we may not be able to select the samples randomly from the population, but at least we can assign subjects randomly to the groups receiving different treatment conditions.

The measurements obtained from random samples have certain characteristics that affect the inferences we make about population values. First, since members have been randomly selected for the samples, they undoubtedly differ from sample to sample; therefore, the measurements (data) obtained from different samples from the same population will also differ.

Second, large samples tend to be more representative of a population than small samples. Therefore, statistics obtained from large samples are more likely to correspond to the values in the population than are small sample statistics. The rest of our study of statistics will be concerned with examining ways of dealing with these two characteristics of measurements from random samples.

We need a way to distinguish between sample and population characteristics. A sample characteristic is called a *statistic*; a population characteristic is called a *parameter*. Thus the mean of a sample is a statistic, whereas the mean of a population is a parameter.

There is a symbol system that always tells us whether we are dealing with statistics or parameters. Up to now, we have considered the small quantities of measurements in our examples to be sample data. Generally, Roman letters are used to denote sample statistics. We have used $\bar{X}$ for the sample mean. When we refer to population parameters, we use Greek letters. The symbol for the population mean is μ (pronounced "mew").

STATISTIC
A characteristic of a sample. Represented by Roman letters.

PARAMETER
A characteristic of a population. Represented by Greek letters.

Recall that Formula 1 gave the formula for calculating a sample mean (symbol $\bar{X}$). If the data for an entire population were available, the formula for its mean would be as given in Formula 2.

FORMULA 2

Calculation of the population mean.

$$\mu = \frac{\Sigma X}{N}$$

Statisticians are seldom privileged to work with data from total populations. More often we are faced with computing statistics from sample data and then making inferences about population parameters. For this reason, the formulas used here are primarily those that are applicable to sample data. We attempt to make estimates of population parameters from the obtained sample statistics. This is the basis for what we term statistical inference.

For example, if we select a random sample of 100 college students and discover that their mean height is 71 inches, we may make the inference that the mean height of the population of college students is also 71 inches. In statistical terms, we say that for $N = 100$, $\bar{X} = 71$. From this we "infer" that $\mu = 71$, knowing that we may be inaccurate in this inference due to sampling error. In fact, estimating the amount of this error is one of the tasks of the statistician.

EXERCISES

1. What does the symbol μ represent?
2. What are sample characteristics called?
3. What are population characteristics called?
4. Are Greek or Roman letters used to represent statistics?
5. Are Greek or Roman letters used to represent parameters?
6. What is meant by a random sample?

6

MEASURES OF VARIABILITY

If we know that a set of data has a mean of 40 or a median of 64, we have representative values for these data, but we still know nothing about the spread of the scores or the extent to which they vary. Bar histograms, frequency polygons, and "smoothed" curves give us graphic representations of the shape of a frequency distribution, but they do not provide us with a statistical way of describing the variability of the scores. To describe a set of data adequately, we need both a measure of central tendency (the mean, the median, or the mode) and a measure of variability. Such measures give us a direct sense of the "differences" on a variable in a given set of data. This chapter explores several ways to describe statistically the variation or dispersion of data.

THE RANGE

The simplest way to represent the spread of the scores in a frequency distribution is to determine the *range*. Formula 3 is the formula for calculating the range of scores in a distribution. It merely determines the difference between the highest score value and the lowest.

FORMULA 3

Calculation of the range.

$$\text{Range} = H - L$$

where: H = highest score in frequency distribution
L = lowest score in frequency distribution

In the frequency distribution shown in Table 6-1, the range of scores is $40 - 32 = 8$. Obviously the range gives us the span of score values in a distribution but tells us nothing about the nature of the distribution of the scores within the limits of the range. The magnitude of the range is affected by only the two extreme values in the distribution.

TABLE 6-1. A frequency distribution of continuous data

X	f
40	2
39	0
38	3
37	9
36	12
35	9
34	8
33	5
32	2
	$N = 50$

THE AVERAGE DEVIATION

The *average deviation*, or *mean absolute deviation*, although rarely used, is a more refined measure of variability than the range because it is computed by using the actual value of each of the scores rather than just its relative position in the distribution. We present it here merely to introduce the concept of variability.

The average deviation is determined by calculating the average of the amounts that the scores deviate from the mean, regardless of the signs of the deviations. To determine the average deviation of the scores from the mean, we must first determine how much each score deviates from the mean. In Formula 4, the symbol x represents the amount of deviation of a score from the mean. It is important here and in future calculations to distinguish between X, which is a "raw" score, and x, which is a deviation score.

FORMULA 4

Calculation of a deviation score.
$$x = X - \bar{X}$$

Formula 5 shows the calculation of the average deviation.

FORMULA 5

Calculation of the average deviation.
$$\text{A.D.} = \frac{\Sigma |x|}{N}$$

where: $|x|$ = the absolute deviation of a raw score from the mean

The term absolute deviation indicates that we are concerned only with *how much* a score deviates from its mean, not with whether it is above or below the mean. Therefore, in determining an absolute deviation score, we ignore the sign of the deviation.

Table 6-2 shows the calculation of the average deviation for the data given in Table 6-1. The mean of these data is $\bar{X} = 35.5$. The deviation of each score value has been determined by Formula 4. The absolute deviation score values are given in the column headed $|x|$. The average deviation of this distribution is 1.44. This tells us that the average number of score points that the scores in the distribution deviate from the mean is slightly less than $1\frac{1}{2}$ points. Thus, unlike the range, the average deviation takes into account the numerical values of the raw scores. However, the average deviation is rarely used in statistical work, because it does not lend itself to further statistical analysis. It is merely a descriptive statistic, which we will not use further in this text.

TABLE 6-2. Calculation of the average deviation

X	f	fX	x	$|x|$	$f|x|$
40	2	80	4.5	4.5	9.0
39	0	0	3.5	3.5	0
38	3	114	2.5	2.5	7.5
37	9	333	1.5	1.5	13.5
36	12	432	.5	.5	6.0
35	9	315	−.5	.5	4.5
34	8	272	−1.5	1.5	12.0
33	5	165	−2.5	2.5	12.5
32	2	64	−3.5	3.5	7.0
	$N = 50$	$\Sigma X = 1775$			$\Sigma|x| = 72.0$

Using Formula 1: $\bar{X} = \dfrac{\Sigma X}{N} = \dfrac{1775}{50} = 35.5$

Using Formula 5: $A.D. = \dfrac{\Sigma|x|}{N} = \dfrac{72}{50} = 1.44$

THE VARIANCE

Probably the two most important measures of variability are the *variance* and the *standard deviation*, because they provide measures that are used in making statistical inferences. We will consider the variance first.

We must introduce two formulas for the variance—one for the calculation of the variance when the total population data are available and one when we are estimating the population variance because we have only a sample of data from the population. We rarely have the total population data for our calculations, but if we did, we would use the following formula for the population variance.

FORMULA 6 Calculation of the population variance.

$$\sigma^2 = \frac{\Sigma x^2}{N}$$

Recall that the symbol x represents the deviation score. Calculating the variance requires squaring each deviation score, summing them, and dividing the sum by N, the size of the population. Note that the variance is denoted by the Greek symbol σ^2, indicating that it is a population parameter. The numerator term Σx^2 is called the "sum of squares," which represents the sum of the squared deviation scores.

FORMULA 7 Calculation of the sum of squared deviations ("sum of squares"). (Formulas 7a and 7b are equivalent.)

Deviation Score Method $\quad \Sigma x^2 = \Sigma(X - \bar{X})^2 \quad$ (Formula 7a)

Raw Score Method $\quad \Sigma x^2 = \Sigma X^2 - \dfrac{(\Sigma X)^2}{N} \quad$ (Formula 7b)

Formula 7 presents two ways to calculate the sum of squares. Formula 7a gives the deviation score method for calculating the sum of squares, and Formula 7b gives the raw score method. These two formulas are algebraically equivalent and they yield identical values. The raw score formula is easier to use when values are to be computed directly from the data. (Formula 6 should be used in those rare instances where data from a total population are available.)

Of more practical value to a statistician is the formula for estimating the population variance when only sample data are available. In estimating the value of the population variance σ^2 from sample data, our first inclination is to use Formula 6 with sample data, and to let this serve as our best estimate of the value of the population variance. However, mathematicians have determined that this would yield a biased estimate of σ^2, since it tends to underestimate the value of σ^2.

Our desire, of course, is to make an unbiased estimate of σ^2. What is an unbiased estimate? Basically <u>we obtain an unbiased estimate of a parameter when the mean of the estimates made from all possible samples of the same size equals the parameter</u>. If we take an infinite number of samples of the same size from a population, however, the mean of all their variances will be slightly smaller than the population variance. For this reason, we say that using Formula 6 with sample data would provide a biased estimate of σ^2, that is, its value, when calculated from the data in a sample, tends to underestimate the population variance. The reason for this is that we would be calculating it by using $\bar{X}$, rather than μ (which is, of course, unknown), as the point of origin for deviation scores. Since $\bar{X}$ undoubtedly is not exactly equal to μ, we have calculated the deviation scores, and hence the variance, from the mean of the sample distribution which yields the smallest possible sum of squares, rather than from μ, which would have given us a larger sum of squares and, consequently, a larger variance estimate.

The question confronting us is: "How do we make an unbiased estimate of σ^2 when we have only sample data?" We can calculate it by using Formula 7 to compute the "sum of squares" and using Formula 8 to compute the variance estimate. Formula 8 gives the formula for estimating a population variance from sample data. We refer to this as simply the sample variance, realizing that it is an estimate of a parameter.

FORMULA 8

Sample variance. Estimate of the population variance from sample data. (Formulas 8a and 8b are equivalent.)

Deviation Score Method $\quad s^2 = \dfrac{\Sigma x^2}{N - 1} \quad$ (Formula 8a)

Raw Score Method $\quad s^2 = \dfrac{N\Sigma X^2 - (\Sigma X)^2}{N(N - 1)} \quad$ (Formula 8b)

where N is sample size.

Formula 8 shows two methods for estimating the population variance from sample data: the deviation score method and the raw score method. These formulas use the Roman symbol s^2 to indicate that it is a sample statistic. Also, the denominator in Formula 8a is $N - 1$, rather than just N as in Formula 6. The denominator of Formula 8b likewise contains $N - 1$. The use of $N - 1$ is necessary to make the estimate an "unbiased" one. Had we used N with sample data, the tendency would have been for the estimate to be somewhat smaller than the true population variance. The subtraction of 1 from the N corrects for this bias.

The reasons for this bias are rather complicated. One explanation is that the size of the variance is greatly affected by large deviation scores. In small samples, it is unlikely that we will obtain extreme scores that may be present in the population. It is more likely that the small number of scores we obtain will be the more typical ones that are not at great distance from the mean. Therefore, because the majority of samples do not contain these extreme scores, using Formula 8 tends to compensate for their absence. Note that this effect is more pronounced in very small samples than in larger ones.

DEGREES OF FREEDOM

In Formula 8, the algebraic expression $N - 1$ is called the number of *degrees of freedom*. The important concept of degrees of freedom pervades much of the remainder of our study of statistics. The degrees of freedom, symbolized as *df*, are determined by the number of scores in a sample minus the number of independent population parameters being estimated. To comprehend thoroughly the meaning of *df*, we need to understand advanced statistical theory, which is beyond the scope of this text. For our purposes, we will concentrate on how the *df* are determined, not on their theoretical underpinnings.

We now have some new terminology to describe the method used in calculating the variance estimate. Instead of reading Formula 8a as: "The population variance is estimated by dividing the sum of squared deviations from the sample mean by the number of observations in the sample minus one," we can now say: "The population variance is estimated by dividing the sum of squares by the degrees of freedom." The latter expression will be of value to us in Chapter 15, which covers analysis of variance.

The number of degrees of freedom is not always determined by $N - 1$. The *df* depend on the number of restrictions that are placed on sample values when a population parameter is estimated. However, the principle is always the same; the *df* of a statistic that is used as an estimate of a parameter are determined by the number of independent values that are used in making the estimate. We will see many uses of the concept of degrees of freedom as we explore various statistical tests.

So that we do not become confused about the symbols for variance and variance estimates that have been used thus far, let's review them.

σ^2 is the variance of a population of measurements.
s^2 is an unbiased estimate of the population variance derived from sample data, called the sample variance.

Let us now apply what we have discussed to a set of sample data. Table 6-3 presents the music aptitude scores of a random sample of 30 music students. Note that Formula 8a and Formula 8b, being algebraically equivalent, yield the same value for the sample variance.

We have discussed the calculation of the variance in detail because it enters into so many of our statistical applications. An examination of Formula 8a reveals that the larger the deviation of scores from the mean, the larger the value of Σx^2, which, in turn, yields a large variance estimate. This further indicates that the variance is a measure of the variability of the scores within a distribution. The smaller the variance, the less dispersion there is among the data.

As evident in Formula 8, the variance is a measure of dispersion, expressed in "squared" units of measurement. The most useful measure of variability is one in which the dispersion of scores is expressed in the actual units of measurement. If we take the square root of the variance, we obtain a measure of variability

TABLE 6-3. Music aptitude scores for 30 students

X	f	fX	X²	fX²
40	1	40	1600	1600
39	3	117	1521	4563
38	5	190	1444	7220
37	9	333	1369	12321
36	6	216	1296	7776
35	4	140	1225	4900
34	2	68	1156	2312
$N = 30$		$\Sigma X = 1104$		$\Sigma X^2 = 40{,}692$

Using Formula 7b: $\Sigma x^2 = 40{,}692 - \dfrac{(1104)^2}{30} = 64.8$

Using Formula 8a: $s^2 = \dfrac{64.8}{30 - 1} = 2.234$

Using Formula 8b: $s^2 = \dfrac{30(40{,}692) - (1104)^2}{30(30 - 1)} = 2.234$

expressed in the original units of measurement. This square root of the variance is called the *standard deviation*.

FORMULA 9

Calculation of the standard deviation.

Population S.D. $\qquad \sigma = \sqrt{\sigma^2} \qquad$ (Formula 9a)

Sample S.D. $\qquad s = \sqrt{s^2} \qquad$ (Formula 9b)

Formula 9 presents the methods for computing the population standard deviation and the sample standard deviation. Formula 9b for the sample standard deviation is really an estimate of the population standard deviation from sample data, since, you will recall, s^2 itself is considered as an estimate of the population variance. The standard deviation for the data in Table 6-3 is calculated as $s = \sqrt{s^2} = \sqrt{2.234} = 1.495$.

For most of our statistical work, the standard deviation is usually considered the standard unit for describing the degree to which scores in a distribution deviate from the mean.

EXERCISES

1. What do the following symbols represent? $s, \sigma, s^2, \sigma^2, x$
2. A manufacturer measured the tolerance of the springs he manufacturers. He obtained the following tolerance levels.

52	50	48	51
51	52	53	50
54	51	51	51
53	54	54	53
50	52	50	49
49	49	49	
50	50	50	

Determine the following values.
a) The range of tolerance levels.
b) The average deviation.
c) The variance by the deviation method.
d) The variance by the raw score method.
e) The standard deviation.
f) The mean tolerance level.
g) The deviation score for a spring with a 53 tolerance level.
h) The deviation score for a spring with a 49 tolerance level.

3. A coach assigned scores on stamina from a possible -10 to a possible $+10$ to each member of her basketball team. She obtained the following scores.

-2	6	0	3	4	4	3	-2
6	8	0	-5	-2	3	4	6
3	0	8	-2	0	4	-4	0
-4	0	-4	4				

Determine the following values.
a) The range of scores.
b) The average deviation.
c) The variance by the deviation method.
d) The variance by the raw score method.
e) The standard deviation.
f) The mean tolerance level.
g) The deviation score for a player with a $+6$ score.
h) The deviation score for a player with a -1 score.

4. A manufacturer wanted to know about the variability in the weights of ball bearings produced by an automated production machine. He selected a random sample of ball bearings and obtained the following weights in grains.

13	24	23	14	23	17
20	12	15	20	17	20
20	27	19	14	19	23
17	19	13	26	16	19
22	17	22	16	23	19
17	25	16	23	17	20
19	21	18	21	19	20
16	21	15	24	20	24
10	11	10	18	11	18
21	18	24	12	21	21

a) Determine the range of weights.
b) Determine the variance of the weights.

c) Determine the standard deviation of the weights.
d) Determine the mean weight of the ball bearings.
e) Prepare a grouped frequency distribution with an interval width of 3, setting the midpoint of the lowest interval at 11.

5. The following data reflect the noise levels of randomly selected third-grade rooms.

73	70	55	64	76	60	65
67	64	78	86	68	71	57
81	90	55	66	82	75	74
67	76	80	73	88	61	72
73	72	56	73	63	82	69
84	70	77	75	65	68	58
66	66	63	63	71	78	74

a) Determine the range of noise levels.
b) Determine the variance of noise levels.
c) Determine the standard deviation of noise levels.
d) Determine the mean noise level.
e) Prepare a grouped frequency distribution with an interval width of 5, setting the midpoint of the lowest interval at 53.
f) Determine the deviation value for a classroom that has a noise level of 83.
g) Determine the deviation value for a classroom that has a noise level of 55.

THE NORMAL DISTRIBUTION

Frequency distributions come in all shapes and sizes. In Chapter 4, we showed curves representing various "shapes" of frequency distributions. Most important for inferential statistics is the normal probability distribution, generally referred to as the *normal distribution*.

The normal distribution is a mathematical construct; that is, it is derived from mathematical theory and thus does not depict a *real* set of data. However, it is extremely useful as a theoretical model that approximates many distributions found in nature. It is particularly valuable in the study of sampling error, which is covered extensively in the next chapter.

THE NORMAL CURVE

A curve representing a normal distribution is called a *normal curve*. The normal curve is represented, not by one specific shape, but by a family of normal curves, each of which has the same properties. Figure 7-1 shows three normal curves. These curves could represent the IQ scores of eighth-grade females, the heights of all 44-year-old men, or the racing times of all runners in the 100-yard dash. Since we now are interested only in the properties of this curve, the actual frequencies and score values it represents are of no concern to us.

For each curve, the values (X) lie along the horizontal axis. The exact shapes of these normal curves differ, depending on the distances between values along the horizontal axis and/or the distances between frequencies along the vertical axis, but we can consider the characteristics of these normal curves to be identical. For convenience, we refer to the family of normal curves simply as the normal curve.

Several characteristics of the theoretical normal curve are of considerable value to the statistician. One such characteristic is *symmetry*. The curve is bell-shaped; that is, it has the same shape on either side of the center point, with

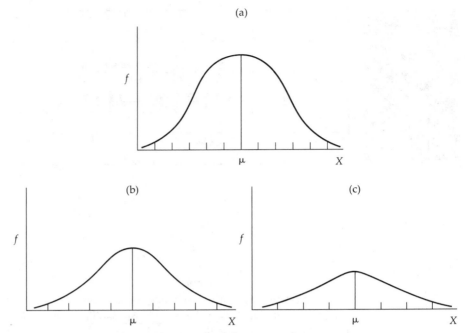

FIGURE 7-1. Three normal curves.

the largest frequency of values located near the center and the smaller frequencies occurring at the two tails of the curve.

Other important characteristics of this curve are that it is *unimodal*, and its mean, median, and mode all coincide at the same value. This means that the mean (arithmetic average) is also the most frequently occurring value (the mode), and it lies at the point that divides the curve exactly in half, with 50% of the population lying above the mean and 50% lying below it (the median).

The curve is also *asymptotic*, in that it extends from the mean toward infinity in both directions. Note that the tails of the curve do not ever actually touch the horizontal axis.

Because it is a theoretical model, the curve is continuous. Thus the distribution of discrete measurements on it only approximates a normal distribution. Conversely, since our actual measurements of continuous variables are always necessarily discrete, they also only approximate this theoretical curve.

Because the theoretical normal curve represents the distribution of an infinite population of measurements, we designate its mean as μ and its standard deviation as σ. The mean is located at the exact center of this curve, as it is in all symmetrical curves. We have introduced the concept of variability, using the variance and standard deviation as measures of dispersion within a set of data. The standard deviation is even more important in statistical work when it is used as a measure of variability of normally distributed data.

We can use the set of normally distributed data that are represented by the curve in Figure 7-2 to illustrate the properties of the normal curve. To show

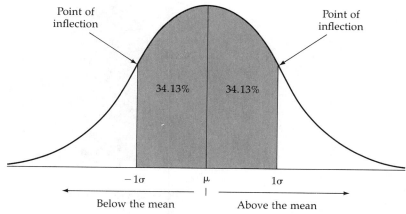

FIGURE 7-2. The normal curve

graphically where the standard deviation is located on a normal curve, we find the point on the curve at which the curve starts growing faster horizontally than it is growing vertically—the point of inflection of the curve—and we draw a perpendicular line from this point to the horizontal axis, as we have done in Figure 7-2. If we were to calculate the value of the standard deviation from the data in this normal distribution, we would find that it does indeed lie at the value on the horizontal axis where the vertical line touches.

Since the curve is symmetrical, what is true of the curve on one side of the mean is also true of it on the other side. Therefore Figure 7-2 shows two points of inflection and two vertical lines dropping to the horizontal axis from these points. The point designated 1σ is said to be one standard deviation above the mean, because it is located on the right of μ, where the values are larger than μ. The same point to the left of μ is designated -1σ and is said to be one standard deviation below the μ.

The entire area under the normal curve, or under any other, is taken as representing 100% of the scores in the distribution. Thus it can be divided into parts representing percentages of the whole. We have already noted that the area to the right of the mean represents 50% of the total area. It is <u>a characteristic of the normal curve that areas containing specific percentages of the total can be determined by using the mean as a point of departure</u>. For example, in Figure 7-2, 34.13% of the total area under the curve lies between the mean and the vertical line we have drawn at 1σ. The distance along the horizontal axis from the mean to the intersection with this line is, by definition, the *standard deviation* of the distribution.

This designation of the standard deviation holds true for both sides of the mean. Thus we have a point on the horizontal axis to the right of μ that is termed one standard deviation above the mean, or simply 1σ, and also a point to the left of μ that is termed one standard deviation below the mean, or -1σ. Both of these points are shown in Figure 7-2. We can see that 68.26%, or about $\frac{2}{3}$, of the scores in the distribution lie between 1σ and -1σ.

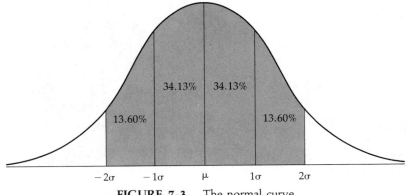

FIGURE 7-3. The normal curve.

If we measure the distance along the horizontal axis from μ to 1σ and then measure the same distance starting at 1σ and moving toward the tail of the curve, we locate the point that represents the second standard deviation, or 2σ. Although the distances along the axis from the mean to 1σ and from 1σ to 2σ are exactly the same, the <u>areas under the curve are different for these two segments.</u> Again, this holds true on both sides of the mean, so we can locate 2σ and -2σ as shown in Figure 7-3. We recall that 34.13% of the area under the curve lies between μ and 1σ. However, it is evident from Figure 7-3 that a much smaller percentage of the area lies between 1σ and 2σ—only 13.60%.

We can now locate a point along the axis that is three times as far from μ as 1σ is; we call this point 3σ. This distance, too, can be measured to the right and to the left of μ, giving us 3σ and -3σ, as shown in Figure 7-4. The figure shows that only a very small percentage of the area under the curve lies between 2σ and 3σ—2.16%, in fact. As we said earlier, since the normal curve is a theoretical curve, it extends to infinity in both directions. The curve virtually

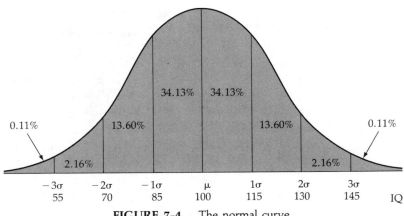

FIGURE 7-4. The normal curve.

touches the horizontal axis, however, at 3σ and -3σ. Indeed, as Figure 7-4 indicates, only 0.11% of the scores lie above 3σ or below -3σ. We can also see in Figure 7-4 that the total area under the curve accounts for 100% of the scores in the distribution (the sum of all the percentages shown).

Now let's examine how to interpret any particular score if we know only how far it is, in standard deviation units, from the mean. To do this, we use a concrete example of IQ scores and assume that they are normally distributed in the population. The curve in Figure 7-4 depicts IQ scores along the horizontal axis. The mean IQ score is 100 and the standard deviation is 15 IQ points; that is, the value of the IQ that lies at 1σ is 15 points larger than the value of μ, or 115. We now have a measure of central tendency, a measure of variability, and the assumption of normality. These characteristics are all we need to obtain a complete picture of the distribution. We know that 50% of the IQ scores are above 100, and 50% are below 100. With $\sigma = 15$, we know that at 1σ the IQ score is 115, and that at -1σ the IQ score is 85. In similar fashion, we can compute the IQ scores at the other σ points shown in Figure 7-4.

Using the percentages associated with the various areas under the curve, we can determine any person's position in relation to the total population by computing how far his IQ score is, in standard deviation units, from the mean of the distribution. For example, suppose a person has an IQ of 130. This is 30 points larger than the mean, or two standard deviations above the mean, and thus falls at the point designated 2σ. We can easily tell that 97.73% of the population have IQs lower than his and that 2.27% have higher IQs.

THE z SCORE

Up to this point, we have found two ways to express a person's score: as a raw score (X) and as a deviation score (x). Another useful measure of a score's position in the distribution, the z score, determines how far a raw score deviates from the mean in standard deviation units. A score expressed in this way is called a *relative deviate*. Formula 10 shows the method for determining the z score corresponding to a particular X.

FORMULA 10

Calculation of a relative deviate (z score).

Population $\quad z = \dfrac{x}{\sigma} \quad$ (Formula 10a)

Sample $\quad z = \dfrac{x}{s} \quad$ (Formula 10b)

Using $\mu = 100$, and $\sigma = 15$, which we have already determined, we can calculate the z score for a person who has an IQ of 130 by first finding the deviation score, $x = 130 - 100 = 30$, and then substituting these values in Formula 10 as follows:

$$z = \frac{30}{15} = 2$$

We could follow this procedure for each score in the distribution. Then each person's IQ score would be expressed in terms of its distance from the mean in standard deviation units. For IQ scores that are larger than the mean, the corresponding z scores are positive. IQ scores that are smaller than the mean have negative corresponding z scores. For example, for IQ = 83, $x = -17$, and $z = -17/15 = -1.13$. This is interpreted as indicating that a person with an IQ of 83 lies at $z = -1.13$ standard deviations from the mean. Thus this person's IQ is expressed in standard deviation terms rather than in deviation score terms.

Note: In statistical work, percentages are less convenient to work with than their decimal equivalents; thus, 34.13% is expressed as .3413 and 13.59% is expressed as .1359. We will use the decimal equivalents from now on.

Figure 7-5 presents the normal curve, and indicates both standard deviations and the proportions of the data that are located within each area. The sum of all the proportions under the curve is 1.0. Note that the horizontal axis is now labeled z, indicating what we learned earlier, that z scores can be used to express raw scores in standard deviation units. When we analyze a distribution of raw scores, we can convert each raw score into a z score using Formula 10, thereby transforming the original distribution into a distribution of z scores. A z score is a standard score, since its mean and standard deviation always have constant values. The mean of a distribution of z scores is always zero and the standard deviation is always 1; z scores are measures of deviation above and below the mean of zero. Therefore, instead of saying: "Jane scored two standard deviations below the mean," we can simply say: "Jane's $z = -2.00$." By examining Figure 7-5, we can immediately tell that .0227 of the population scored below Jane and .9773 scored above her. This illustrates how helpful z scores are when we are dealing with normally distributed data.

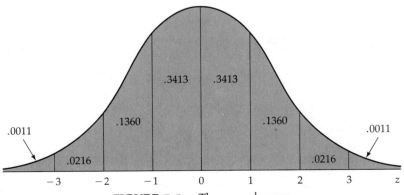

FIGURE 7-5. The normal curve.

Up to now, we have been dealing with areas bounded at specific standard deviation points, or z scores. We need a method for computing the proportions of areas with boundaries falling *between* any two z scores, even when these are fractional values. Table 1, at the end of the book, presents the proportions of the area under the curve associated with different z scores. Using this table, we can determine the proportion of the area under the normal curve corresponding to any interval for which the boundaries are expressed as z scores. The first column in Table 1 presents z-score values, and the second column shows the proportion of the total area under the curve that lies between μ and any particular z score.

To use this table we must first convert the raw scores to z scores. In our example using IQ scores, we would use Formula 4 to determine the deviation score for a person with an IQ of 125. (Here μ is substituted for $\bar{X}$, since we are dealing with a population parameter rather than a sample statistic.)

$$x = X - \mu = 125 - 100 = 25$$

Then, using Formula 10, we determine the z score:

$$z = \frac{x}{\sigma} = \frac{25}{15} = 1.67$$

Therefore a person with an IQ of 125 has a z score of 1.67. Note that the value of the z score represents the distance expressed in the number of standard deviations the raw score lies away from the mean. If the z score is positive,

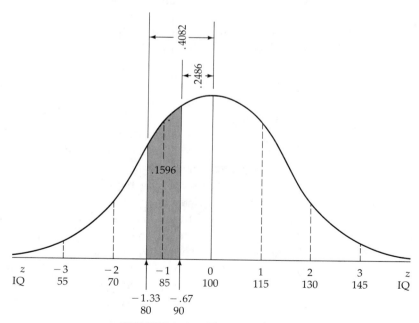

FIGURE 7-6. The normal curve.

it represents a value above the mean; if it is negative, it indicates a value below the mean. Table 1 shows that the proportion of the total area under the curve that lies between μ and $z = 1.67$ is .4525.

In another example using Table 1, suppose we want to find the proportion of the population with IQs between 80 and 90. The z score equivalents for these two IQs are $z = -1.33$ and $z = -.67$, respectively. (Remember, the sign of the z score tells us whether it is located above or below the mean. Table 1 is used for both positive and negative z scores, since the areas under the curve remain the same in both cases.) Table 1 tells us that the proportion of the total area from μ to $z = -1.33$ is .4082 and from μ to $z = -.67$ is .2486. The proportion lying between these two z scores, or between IQs 80 and 90, is $.4082 - .2486 = .1596$. This is shown graphically in Figure 7-6.

In summary, the total area under the normal curve is considered 1.0. The proportion of this area that lies between any two raw score values can be determined by converting the raw scores to z scores, and then using Table 1 to determine the proportions of the total area that lie between the mean and the respective z score values.

EXERCISES

1. A group of elementary school children were given a test designed to measure divergent thinking ability. The children's scores had a mean of 30 points and a variance of 16 points. Assume the scores were normally distributed. Determine:
   a) The standard deviation for these data.
   b) The relative deviate for score 24.
   c) The proportion of children who received scores of 24 or less.
   d) The relative deviate for score 37.
   e) The proportion of children who received scores of 37 or greater.
   f) The proportion who received scores from 25 through 32.
   g) The proportion who received scores from 31 through 33.
   h) The proportion who received scores from 26 through 28.
2. A group of graduates from a business college were given a clerical aptitude test. Their scores were normally distributed, with a mean of 108 and a variance of 400. Determine:
   a) The standard deviation for these data.
   b) The relative deviate for score 129.
   c) The proportion of the graduates who received a score of 129 or greater.
   d) The relative deviate for score 109.
   e) The proportion who received a score of 109 or less.
   f) The proportion who received scores from 70 through 130.
   g) The proportion who received scores from 100 through 120.
   h) The proportion who received scores from 55 through 75.

PROBABILITY

The concept of probability is the basis on which all statistical decisions are made. If we were not blessed with probability theory, statistics would be limited to the rather mundane function of mere description. But through the application of statistical formulas, we can specify the probability that events will occur by chance, and thereby make intelligent judgments about experimental results. The processes underlying statistical inference are based entirely on considerations of probability.

Probabilities range from 0 to 1.00. If we roll a standard pair of dice, the probability that we will roll a 13 on a given roll is 0. The probability that our pet rooster will lay an egg tomorrow is also 0. When it is impossible for an event to occur, its probability is 0.

If we toss a coin, there are two possible outcomes, and the probability that it will land either heads or tails is 1.00. If I release my pencil, the probability that it will fall to the desk is also 1.00. When it is certain that an event will occur, its probability is 1.00.

These examples are unequivocal. When we make statistical inferences in the behavioral sciences, however, we are never in a position to enjoy the luxury of these extremes. We are forced to make interpretations of data in spite of some degree of uncertainty. Therefore we must understand how probability is determined so that we can make judicious use of it.

When we are able to specify all of the possible outcomes that may happen, and to determine the relative frequency with which unique outcomes may occur on the basis of chance alone, we can determine the *probability* that an outcome will occur. (The term *relative frequency* refers to the proportion of times each outcome occurs relative to the total number of possible outcomes.) This probability is stated as a proportion of 1.00.

For example, consider a die with 6 sides, each of which shows a number from 1 through 6. There are six possible outcomes that can occur on one roll of this die. If we rolled it an infinite number of times, the relative frequency

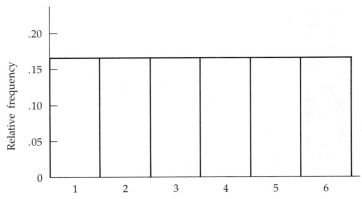

FIGURE 8-1. Relative frequency distribution of outcomes for one die.

of occurrence of each number would be 1/6, given that the die is fair. That is, the probability of rolling a 5 on any one roll would be 1/6 or .167. The relative frequency of occurrence for each number on the die is shown graphically in Figure 8-1. This figure depicts a rectangular distribution, because each possible outcome has an equal probability (.167) of happening.

We call the relative frequency distribution in Figure 8-1 a *probability distribution*. From this distribution, we can see that the probability of rolling either a 1 or a 2 on a particular roll is .333, determined by summing .167 (for number 1) and .167 (for number 2). The probability of rolling a number larger than 3 is .500 (.167 + .167 + .167). Of course, the probability of rolling a number from 1 through 6 is 1.000. In statistics we use the symbol P for probability. Therefore, the probability of rolling a 2, 4, 5, or 6 on one roll of the die is $P = .667$.

If a *pair* of dice is rolled an infinite number of times, there are 36 possible combinations of dice surfaces, each of which yields 11 possible outcomes—that is, total points. The relative frequency of occurrence for each of these possible outcomes is shown in Table 8-1. This probability distribution of throws of a pair of dice is diagrammed in Figure 8-2.

Two of the points just made should be emphasized. First, we must note that probability distributions occur as a result of chance; that is, they come about when randomness is free to have its way over an infinite number of events. Second, we have used the expression *will occur* in connection with the concept of probability. This expression refers to a speculation about the relative frequency with which a specific event will take place, based on a given probability distribution. In inferential statistics we can also reverse this process; when we obtain a sample statistic, we can use an appropriate probability distribution to answer the question: "What is the probability that the statistic we obtained could have happened by chance?" Answering this question is one of the major functions of statistical analysis.

TABLE 8-1. Relative frequency of occurrence

Total points	Possible ways of occurring	Relative frequency
2	1	.028
3	2	.056
4	3	.083
5	4	.111
6	5	.139
7	6	.167
8	5	.139
9	4	.111
10	3	.083
11	2	.056
12	1	.028
	36	

The probability distributions given in Figures 8-1 and 8-2 depict relative frequencies for discrete data. (A pair of dice can come up totaling only 11 discrete numbers.) In such cases, it is possible to specify the probability of each occurrence. This is not so when continuous variables are involved. This is because when we work with continuous variables, we are theoretically dealing with an infinite number of different values under the curve. It would be impossible to divide the total area, which is considered to be 1.00, by an infinite number of possible values. However, we can compute probabilities for areas between any two values under the curve. Thus probabilities in continuous probability distributions are defined in terms of intervals under the curve rather than in terms of individual values.

Let's apply this discussion of probability to the normal curve. We may consider the normal curve to be a probability distribution that has been derived

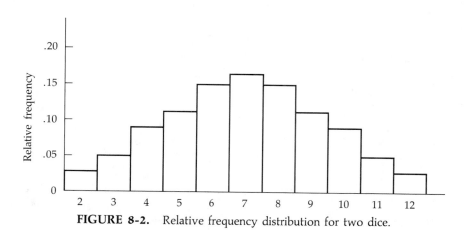

FIGURE 8-2. Relative frequency distribution for two dice.

from mathematical theory. This is one of the most useful probability distributions in the study of statistics. Since it is a distribution of a continuous variable, we can only specify probabilities for areas under it.

The probability of randomly selecting a score from a given area under the curve is equal to the proportion of the total number of scores that is contained in the area. Therefore, when the normal curve represents a probability distribution, the probability that a score will be randomly selected from a particular area can be determined through the use of z scores and reference to Table 1. This means that the proportions that we calculated in Chapter 7 can be considered probability statements and the values given in Table 1 can be interpreted as probabilities.

We can answer the question: "What is the probability that a randomly selected individual represented in Figure 7-4 will have an IQ between 80 and 90?" in the following manner. We merely convert the proportion of .1596 into a probability statement: "The probability that we would select at random from this population an individual who has an IQ between 80 and 90 is $P = .1596$." In other words, about 16 times out of 100 (or about 16% of the time) we would randomly select an individual who has an IQ within this range. We can also state that the probability of randomly selecting an individual with an IQ below 70 is $P = .0227$; therefore, it is a rather unlikely occurrence.

We can use Table 1 to determine the probability of selecting at random a person who has a z score of, for example, -1.34 or lower. Table 1 shows the probability of obtaining a score lying between μ and $z = -1.34$ to be $P = .4099$. We know that for values below μ, $P = .5000$. Therefore, the probability of obtaining a z score of -1.34 or lower is $P = .5000 - .4099 = .0901$.

Let's look at one more example of the procedure for determining probabilities using Table 1. Suppose we know that the population of completion times for a test of dexterity is normally distributed, with $\mu = 45$ minutes and $\sigma = 5$ minutes. What is the probability that a randomly selected individual will complete the test in from 38 to 47 minutes? Because time is considered a continuous variable, we must use the limits of 38 and 47 in our calculations. The limits of 38 are 37.5–38.5; the limits of 47 are 46.5–47.5. Therefore, we are interested in the interval from 37.5 to 47.5.

We must first determine the z scores for these two values and their associated probabilities from Table 1.

For $X = 38$: $\quad z = \dfrac{x}{\sigma} = \dfrac{37.5 - 45}{5} = -1.5. \quad$ P between μ and $z = .4332$

For $X = 47$: $\quad z = \dfrac{x}{\sigma} = \dfrac{47.5 - 45}{5} = 0.5. \quad$ P between μ and $z = .1915$

The probability of selecting an individual with a test time that falls between these two z scores is $P = .4332 + .1915 = .6247$. This example is diagrammed in Figure 8-3.

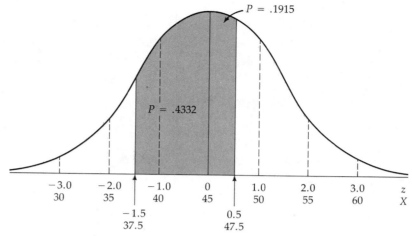

FIGURE 8-3. The normal curve.

The important point to remember when using Table 1 is that the probabilities given in the table are for the area between the mean and the z scores, whether the z scores are positive or negative.

EXERCISES

1. What does the symbol P represent?
2. What is another name for a relative frequency distribution?
3. For a population of normally distributed data with $\mu = 40$ and $\sigma = 4$, use Table 1 to determine:
    a) The probability of randomly selecting a score between μ and 46.
    b) The probability of selecting a score between μ and 31.
    c) The probability of selecting a score between 35 and 45.
    d) The probability of selecting a score as large as or larger than 47.
    e) The probability of selecting a score as small as or smaller than 32.

9 THE DISTRIBUTION OF SAMPLE MEANS

We have by now interpreted the area under the normal curve in terms of percentages, proportions, and probabilities. In our previous examples, the curve has represented the distribution of the individual measures comprising the population. We now extend this use of the curve to the calculation of means of samples.

When we randomly select samples from a population, it is highly unlikely that the value of each sample mean will equal the value of the population mean, since very few samples are exact microcosms of the population. Completely by chance, some of the samples will have $\bar{X}$s larger than μ, and some will have $\bar{X}$s smaller than μ. Many of them will have $\bar{X}$s close to μ, and few will deviate greatly from μ on either side. We need to know how the means of randomly selected samples are distributed.

The following three theorems, which are derived from the Central Limit Theorem, are the foundations on which lies much of our ability to draw inferences from sample data.

THEOREM 1 *The means of a multitude of equal-sized samples drawn from a normally distributed population are themselves normally distributed.*

THEOREM 2 *The means of a multitude of equal-sized samples, regardless of the shape of the population distribution, approach a normal distribution as the sample size increases.*

THEOREM 3 *The mean of a multitude of normally distributed sample means is the population mean.*

Theorem 1 holds true regardless of the number of observations in each sample. Theorem 2 is important because, in the behavioral sciences, we are generally concerned with variables for which little or nothing is known about the form of the population distribution. In general, a sample size of 30 or more is considered large enough for us to use the normal curve as a close approximation of the distribution of sample means. Theorem 3 tells us that the value of the mean of all the sample means is equal to the value of the mean of the population from which the samples were drawn. The probability distribution of the sample means is called a *sampling distribution of means*, with the population mean at the center of the distribution.

Of what practical value is it to know that the distribution of means of samples, each based on 30 or more observations, closely resembles the normal curve? It signifies that the standard deviation of the sample means can be used in conjunction with Table 1 to determine probabilities for the means, just as we did for individual scores. The standard deviation of the means of a multitude of equal-sized samples is called the *standard error of the mean*, and the symbol for this parameter is $\sigma_{\bar{X}}$.

Figure 9-1 presents a distribution of sample means of reading scores, where each sample has $N = 50$. Theorems 2 and 3 hold for this example. (Remember that this is a distribution of mean scores of a large number of samples, not a distribution of individual scores.) The curve in Figure 9-1 is a normal curve, and all of the properties we have attributed to it in connection with the distribution of individual scores are also valid for the distribution of sample means. We can see from Figure 9-1 that the mean of all the sample means is μ. Remember that these samples are all of the same size.

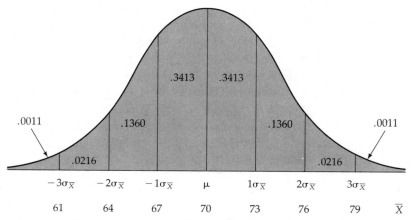

FIGURE 9-1. Sampling distribution of mean reading scores.

In Figure 9-1, the σ units are designated as $\sigma_{\bar{X}}$ because this is a distribution of sample means. In this example, we know that the population mean reading score is 70, and we can determine the $\sigma_{\bar{X}}$ from the distribution of sample means around μ by computing the standard deviation of the sample means. (An alternate way to calculate $\sigma_{\bar{X}}$ is to use Formula 11a, to be presented shortly.) Let's suppose that our calculations yield $\sigma_{\bar{X}} = 3$. This tells us that the standard error of the mean is three points. Knowing this, and using Table 1, we can then determine the probability of selecting a sample with a mean within any particular interval of mean scores.

The probability of obtaining a sample with a mean between 68 and 75 can now be determined for this example. Since reading achievement is considered a continuous variable, the real limits of this interval are 67.5 and 75.5.

For $\bar{X} = 67.5$: $z = \dfrac{67.5 - 70}{3} = -0.83$. From μ to z, $P = .2967$.

For $\bar{X} = 75.5$: $z = \dfrac{75.5 - 70}{3} = 1.83$. From μ to z, $P = .4664$.

Therefore, the probability of obtaining from this population a sample of 50 individuals that has a mean reading score between 68 and 75 is $.2967 + .4664 = .7631$.

This use of the normal curve as an approximation of the sampling distribution of means forms the basis for some of the major statistical inference procedures.

Up to this point, we have occasionally indulged in the "conceptual luxuries" of dealing with populations in which parameters were known; of taking infinite numbers of samples from such populations and examining the distributions of their means; and of determining probabilities in selecting samples, given a known sampling distribution. These examples helped to show how data "behave" in populations and samples.

Now we turn to more practical applications of our statistical knowledge. As has been said earlier, very seldom do we have access to the values of population parameters. Our efforts are usually directed toward making estimates of population parameters from sample data. Recall that Formulas 8 and 9b have given us ways of estimating the variance and standard deviation from data in one sample.

When we deal with the relationship between a population and its samples, we must be aware of two statistical facts:

1. A population has parameters with specific values (these are usually not known)—that is, in a given population, the mean has one value which does not vary. The variance of all measurements in a population is also a specific value.
2. Sample statistics, such as $\bar{X}$ and s^2, vary from sample to sample because of *sampling error*.

These two facts lead us to our first problem in statistical inference. Given the data in a sample, what is our best estimate of the parameters of the population from which it was selected?

ESTIMATING THE STANDARD ERROR OF THE MEAN

In Figure 9-1, we illustrated how a sampling distribution of sample means can be developed around μ by computing the standard deviation of a multitude of sample means of like-sized samples. Recall that this standard deviation is called the standard error of the mean and is symbolized by $\sigma_{\bar{X}}$. But how can we develop such a sampling distribution when we cannot obtain a multitude of samples? Fortunately statisticians have developed a formula for estimating the value of the standard error of the mean using only the data in one sample. This estimate is symbolized by $s_{\bar{X}}$, because it is derived from sample data.

FORMULA 11

Calculation of the standard error of the mean.

Population $$\sigma_{\bar{X}} = \frac{\sigma}{\sqrt{N}}$$ (Formula 11a)

Sample $$s_{\bar{X}} = \frac{s}{\sqrt{N}}$$ (Formula 11b)

Formula 11a presents the method for calculating the standard error of the mean where the standard deviation of a population is known. Formula 11b presents the method for estimating the standard error of the mean from sample data. In both formulas, N is the size of the sample.

Formula 11b shows that dividing the estimate of the population's standard deviation by the square root of the sample size yields an estimate of the standard deviation of sample means, which is called the *estimate of the standard error of the mean*. This formula is extremely valuable because it allows us to estimate from the data in a single sample how a multitude of sample means would vary around the population mean.

Examination of Formula 11b reveals that the size of $s_{\bar{X}}$ is a function of two values, s and N. For samples of the same size, the larger the s is, the larger $s_{\bar{X}}$ will be. This is logical because the means of random samples would be expected to vary more in populations with widely varying scores than in populations with relatively homogeneous scores.

On the other hand, for a population with a given s, the larger the sample is, the smaller the $s_{\bar{X}}$ will be. This also stands to reason because we would expect the means of large samples to vary less from μ (have less error) than the means of small samples. We always wish to reduce error, and now we have discovered that one way to accomplish this is to obtain large samples.

Table 9-1 gives concise examples of the methods of calculation introduced in this chapter. This example shows the computation of the standard error of the mean from the data in one sample, using both the raw-score method and the deviation-score method.

Having estimated the standard error of the mean from sample data, we could develop a sampling distribution of sample means around the population mean. But, alas, we seldom know the value of the population mean. However, we can

TABLE 9-1. Calculation of the estimate of the standard error of the mean

X	f	fX	X²	fX²	x	x²	fx²
14	1	14	196	196	4	16	16
13	3	39	169	507	3	9	27
12	2	24	144	288	2	4	8
11	8	88	121	968	1	1	8
10	7	70	100	700	0	0	0
9	6	54	81	486	−1	1	6
8	3	24	64	192	−2	4	12
7	3	21	49	147	−3	9	27
6	1	6	36	36	−4	16	16
N = 34		ΣX = 340		ΣX² = 3520			Σx² = 120

$$\bar{X} = \frac{340}{34} = 10$$

Calculation of sum of squares:

Formula 7a: $\Sigma x^2 = \Sigma(X - \bar{X})^2 = 120$ (right-hand column)

Formula 7b: $\Sigma x^2 = \Sigma X^2 - \frac{(\Sigma X)^2}{N} = 3520 - \frac{(340)^2}{34} = 120$

Calculation of estimate of population variance:

Formula 8a: $s^2 = \frac{\Sigma x^2}{N - 1} = \frac{120}{34 - 1} = 3.64$

Formula 8b: $s^2 = \frac{N\Sigma X^2 - (\Sigma X)^2}{N(N - 1)} = \frac{34(3520) - (340)^2}{34(34 - 1)} = 3.64$

Calculation of estimate of population standard deviation:

Formula 9b: $s = \sqrt{s^2} = \sqrt{3.64} = 1.91$

Calculation of estimate of the standard error of the mean:

Formula 11b: $s_{\bar{X}} = \frac{s}{\sqrt{N}} = \frac{1.91}{\sqrt{34}} = .33$

---

put the estimate of the standard error of the mean to good use in establishing *confidence intervals* for the value of the population mean. This method is given in the next chapter.

EXERCISES

1. What do the following symbols represent? $\sigma_{\bar{X}}, s_{\bar{X}}$
2. Suppose that a multitude of same-sized samples with $N = 81$ were randomly selected from a population where $\mu = 70$ and $\sigma = 16$.
   a) What is the standard error of the mean?
   b) What is the probability of randomly selecting from this population a sample of 81 people that has a mean score of 74 or greater?
   c) What is the probability of a sample that has a mean of 67 or less?
   d) What is the probability of a sample that has a mean that falls between 66 and 68?

e) What is the probability of a sample that has a mean that falls between 72 and 75?

f) What is the probability of a sample that has a mean that falls between 68 and 72?

3. Suppose that a multitude of same-sized samples with $N = 64$ were randomly selected from a population where $\mu = 120$ and $\sigma = 12$.

   a) What is the standard error of the mean?

   b) What is the probability of randomly selecting from this population a sample of 62 people that has a mean score of 115 or greater?

   c) What is the probability of a sample that has a mean of 120 or less?

   d) What is the probability of a sample that has a mean that falls between 120.5 and 128.7?

   e) What is the probability of a sample that has a mean that falls between 118 and 122?

   f) What is the probability of a sample that has a mean that falls between 116 and 119?

10 ESTABLISHING CONFIDENCE INTERVALS

One of the common statistical problems we face when we are confronted with a set of sample data is how to make an "educated guess" concerning how accurately the value of the sample mean represents the value of the unknown population mean. In other words, we want to use the sample values to provide a basis for an inference about the value of the parameter μ.

As background for this problem in statistical inference, let us first consider a population from which we can select an infinite number of samples of 50 cases each. As we saw in Chapter 9, we can prepare a distribution of sample means. The mean of such a distribution will be the mean of the population, and the standard deviation of the sample means is the standard error of the mean. Since the means of samples are normally distributed, we can use our knowledge of the probabilities associated with the areas under the normal curve as it applies to a distribution of sample means.

Suppose we now want to determine the interval, centering on μ, within which .95 of the sample means lie. To establish this interval we need to specify, in $\sigma_{\bar{X}}$ units (or z-score units), the points above and below μ between which .95 of the sample means are contained. Since Table 1 shows the proportions of the area under the normal curve located between the mean and each z score, we need to find the z score that demarks .4750 of the total area above μ and .4750 of the total area below μ. From the "μ to z" column of Table 1, we find that .4750 is associated with the z score 1.96. Therefore, the boundaries of the interval containing .95 of the sample means are located at $z = -1.96$ and $z = 1.96$.

The interval bounded by these two points and the relevant probability areas are shown in Figure 10-1. This figure shows that .95 of the sample means lie between $-1.96\sigma_{\bar{X}}$ and $1.96\sigma_{\bar{X}}$. We already know that this proportion can be considered a probability. Therefore, if we select one sample from this distribution at random, the probability is .025 that its mean lies below $-1.96\sigma_{\bar{X}}$ and

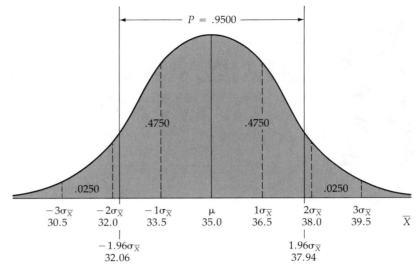

FIGURE 10-1. Interval containing 95% of the sample means.

.025 that it lies above $1.96\sigma_{\bar{X}}$. On the other hand, the probability of our selecting a sample with a mean that lies within the interval bounded by $-1.96\sigma_{\bar{X}}$ and $1.96\sigma_{\bar{X}}$ is $P = .95$.

For example, suppose we know that the mean of a population is 35 and $\sigma_{\bar{X}}$ is 1.5. To determine the interval containing .95 of the $\bar{X}$s, we must determine the values of the sample means at $-1.96\sigma_{\bar{X}}$ and $1.96\sigma_{\bar{X}}$. For $-1.96\sigma_{\bar{X}}$, this value lies at 1.96×1.5 below the mean, which is $35 - 2.94$, or 32.06. For $1.96\sigma_{\bar{X}}$, this value lies at 1.96×1.5 above the mean, which is $35 + 2.94$, or 37.94. This interval is shown in Figure 10-1.

We can say that there is a probability of .95 that the mean of one sample randomly selected from this distribution will lie between 32.06 and 37.94. We can also say that the probability that its mean will have a value outside (either above or below) this interval is .05.

The logic used in these examples will help us to understand how the properties of the normal curve can be used to establish an interval within which there is a specified probability of an $\bar{X}$ occurring. However, since we are rarely in a position to take an infinite number of samples from a population, let us now return to a more typical case, in which we have only one sample and must use the data obtained from it to make inferences about the population mean. For example, suppose we want to estimate the mean height of males in the United States. If we could measure all men in the nation, and calculate their mean height to be 69 inches, we could state with certainty, barring measurement errors, that this represents the mean height of American males. But if we can take only one random sample of males and compute the mean height of this sample, we cannot state with complete confidence that the mean of the sample is identical with the population mean. This is true because, as we have seen, the means of samples vary around the population mean; few samples are likely to yield means that are identical with the population mean.

Of course, the sample mean is the one value that is our best estimate of the population mean, in cases where we have data from only one sample. Suppose we randomly select a sample of males and calculate their mean height to be 68 inches. Then our best guess for the population mean has to be 68 inches. However, we will have little confidence that our guess is correct, since we know that a sample mean is likely to be located either above or below the unknown population mean. When we use the sample mean as an estimate of the population mean, we are making a *point estimate* of the population value. It is difficult to know how much confidence to place in this point estimate, because we have no information by which to gauge its accuracy.

We need a procedure that will enable us to state the degree of confidence we have in an estimate. One such procedure entails establishing a *range* of values for estimating the population mean, instead of just taking the sample mean as a point estimate of it. If we make an *interval estimate* using sample data, we can determine the degree of confidence we have that our interval contains the population mean. Our procedure, then, is to establish a *confidence interval* from sample data from which we can make inferences about the population mean.

THE 95% CONFIDENCE INTERVAL

One commonly used method for making statements about the value of the population mean is to determine the 95% confidence interval. By using this procedure, we determine an interval such that, if we repeated the procedure for a large number of samples, 95% of them would encompass the population mean.

Our method for establishing the 95% confidence interval follows. We have shown that the probability is .95 that a randomly selected z score lies between -1.96 and 1.96. This can be shown as

$$P(-1.96 \leq z \leq 1.96) = .95$$

This expression is read: "The probability that a z score is equal to or greater than -1.96 and is equal to or less than 1.96 is .95."

This expression can also be applied to the deviation of sample means from the population mean. To do this, we define z as $(\bar{X} - \mu)/\sigma_{\bar{X}}$. Substituting this in the preceding expression, we have

$$P\left(-1.96 \leq \frac{\bar{X} - \mu}{\sigma_{\bar{X}}} \leq 1.96\right) = .95$$

Now, multiplying through by $s_{\bar{X}}$, this expression becomes

$$P(-1.96\sigma_{\bar{X}} \leq \bar{X} - \mu \leq 1.96\sigma_{\bar{X}}) = .95$$

If we subtract $\bar{X}$ from each member of the expression, we have

$$P(-\bar{X} - 1.96\sigma_{\bar{X}} \leq -\mu \leq -\bar{X} + 1.96\sigma_{\bar{X}}) = .95$$

Changing signs throughout the expression, which also reverses the sign of the inequalities, we have

$$P(\bar{X} + 1.96\sigma_{\bar{X}} \geq \mu \geq \bar{X} - 1.96\sigma_{\bar{X}}) = .95$$

Rearranging the terms gives

$$P(\bar{X} - 1.96\sigma_{\bar{X}} \leq \mu \leq \bar{X} + 1.96\sigma_{\bar{X}}) = .95$$

This last expression gives us the specification of the 95% confidence interval. It is read as: "The probability is .95 that the interval bounded by $\bar{X} + 1.96\sigma_{\bar{X}}$ and $\bar{X} - 1.96\sigma_{\bar{X}}$ encompasses the population mean."

The development of the preceding expressions made use of $\sigma_{\bar{X}}$, which, in reality, is never directly available to us. Instead, we estimate its value from sample data using Formula 11b, giving us $s_{\bar{X}}$. Replacing $\sigma_{\bar{X}}$ with its estimate, $s_{\bar{X}}$, in the preceding expression we obtain

$$P(\bar{X} - 1.96s_{\bar{X}} \leq \mu \leq \bar{X} + 1.96s_{\bar{X}}) = .95$$

The preceding expression holds true for establishing the 95% confidence interval for relatively large samples, say, where $N = 100$ or more, but yields inaccurate intervals with smaller samples. A method that can be used with small samples is presented later.

To illustrate how the confidence interval is established, we randomly select a sample of 100 American males and measure their heights. Assume that in the sample, $\bar{X} = 68$ inches, and $\underline{s^2 = 36}$. The point estimate of μ would then be 68 inches. We know, however, that this is a more or less inaccurate estimate; a confidence interval will give us more useful information. Therefore, using Formula 11b, we estimate that $s_{\bar{X}} = .6$ inches.

We have shown that the 95% confidence interval is determined by the value that lies at $-1.96s_{\bar{X}}$ and the value that lies at $1.96s_{\bar{X}}$. If $\bar{X} = 68$ and $s_{\bar{X}} = .6$, then the value that lies at $-1.96s_{\bar{X}}$ must be $-1.96 \times .6$, or -1.18 inches from the mean. This value, which represents the lower boundary of the 95% confidence interval, is 66.82 inches.

The upper boundary of the interval is $1.96 \times .6$ or 1.18 inches from the mean, which is 69.18 inches. Therefore the 95% confidence interval developed from the data in this sample is from 66.82 to 69.18 inches.

We reason that if we could obtain a large number of samples, each with $N = 100$, and determine an interval from each of them using this method, 95% of these intervals would encompass the population mean. Since we have used the data from only one sample to establish the confidence interval, we conclude that the probability is .95 that our interval is one of those that encompasses the mean height of the male population.

Of course, it is obvious that the μ must have a specific value, although it is not known to us. Either the value of μ is within our established 95% confidence interval, or it is not. Because it is an all-or-nothing situation, the concept of probability is not involved. Thus it is actually incorrect to state that the probability that μ lies within the interval is .95.

How do we interpret the confidence interval? If we took a multitude of samples and calculated the 95% confidence interval for each sample, we would find that, because of sampling error, the limits of the intervals would vary from sample to sample. Some of these intervals, 95% of them, in fact, would contain μ and some would not. This provides a basis for interpreting the meaning of a confidence interval. But in this case, we do not have a multitude of samples; we have only one sample, from which we have developed one 95% confidence interval. We cannot say that the probability is .95 that μ lies within our interval. We can say, however, that <u>if this process of establishing intervals is applied to many samples, it will yield intervals that contain the population mean 95 times out of 100</u>. We must t<u>herefore conclude that the probability is .95 that our interval is one of those that encompasses μ</u>.

THE 99% CONFIDENCE INTERVAL

If we need an interval that will have an even greater probability of containing μ, we can calculate the 99% confidence interval. We can then state that, in the long run, 99% of the intervals determined by this method will encompass μ.

The method used to determine the 99% confidence interval is similar to the procedure for calculating the 95% confidence interval. As we would expect, the 99% confidence interval will be wider than the 95% confidence interval for the same data. But we will have a higher degree of confidence that the 99% interval encompasses μ.

The first step is to determine the two points on the horizontal axis of the normal curve that designate the boundaries of the central .99 of the area under the curve. Table 1 shows that .4951 of the area lies between μ and $z = 2.58$. Since we are indicating variability in terms of $s_{\bar{X}}$, we can say that .99 of the area is contained within the interval $-2.58s_{\bar{X}}$ to $2.58s_{\bar{X}}$.

In the example used earlier, in which $\bar{X} = 68$ inches and $s_{\bar{X}} = .6$ inches, the boundaries of the 99% confidence interval are determined as follows. The value at $-2.58s_{\bar{X}}$ lies at $-2.58 \times .6$ from the mean, which is $68 - 1.55$, or 66.45 inches. The value at $2.58s_{\bar{X}}$ lies at $2.58 \times .6$ from the mean, or 69.55 inches. Note that the 99% confidence interval is wider than the 95% confidence interval, which covered the area from 66.82 to 69.18. Intervals formed in this manner for a multitude of samples would encompass the μ 99% of the time. Thus probability that our interval based on one sample is one of these encompassing μ is $P = .99$.

In summary, the probability expressed in a confidence interval refers to the likelihood that it is one of the many possible intervals that can be determined for one population that contain the population mean. It does not state the likelihood that the population mean will fall within a given interval. There is only one population mean with a specific value (although it is unknown to us), but there are as many possible intervals as there are possible samples. Each sample yields its own interval, and these intervals differ from sample to sample. In computing a confidence interval for a given sample, we are attempting to determine the probability that it is one of those intervals that encompass the population mean, rather than one of those that do not.

We could, of course, determine any other confidence interval for the data in our sample. For example, using Table 1, we could compute the 50% confidence interval or the 80% confidence interval. We have used the 95% and 99% confidence intervals here because they are the ones most commonly used in making inferences about the population mean.

THE t DISTRIBUTION

Up to this point, we have considered two situations in which the population values were assumed to be normally distributed:

1. In cases where the parameter σ was known, we determined $\sigma_{\bar{X}}$ by $\sigma/\sqrt{N}$ and found that the z scores calculated by $(\bar{X} - \mu)/\sigma_{\bar{X}}$ were normally distributed, with a mean of 0 and a standard deviation of 1, regardless of the sample size.
2. In cases where σ was unknown, we estimated it by using $s_{\bar{X}} = s/\sqrt{N}$. We found that the ratio $(X - \mu)/s_{\bar{X}}$ was approximately normally distributed for large-sized samples.

The first procedure can be used with both large- and small-sized samples. However, its usefulness is limited since we seldom know the value of σ. The second procedure is useful when we have relatively large samples (those with an N of 30 or more), but as the sample size becomes smaller, the distribution of the ratio, $(\bar{X} - \mu)/s_{\bar{X}}$, becomes increasingly dissimilar to the normal curve.

We must now learn how to establish the sampling distribution of $(\bar{X} - \mu)/N$ for small samples when the value of $\sigma_{\bar{X}}$ is unknown. In such instances, we estimate the standard error of the mean using Formula 11b. Since $s_{\bar{X}}$ is calculated from the data in a sample, we know that its value will vary from sample to sample. We also know that there will be less variability in the values of $s_{\bar{X}}$ when we use a multitude of large-sized samples than there will be when we use small-sized samples—that is, the smaller the sample size, the more variability there will be in $s_{\bar{X}}$ from sample to sample. Thus we must conclude that the $s_{\bar{X}}$ derived from a large sample is a closer approximation of $\sigma_{\bar{X}}$ than is one derived from a small sample.

Let's examine more closely what this means. We have stated that z scores computed from the ratio $(\bar{X} - \mu)/\sigma_{\bar{X}}$ are normally distributed, regardless of sample size. This is because where σ is known, $\sigma_{\bar{X}}$ can be calculated and it is a fixed value. The μ is also a fixed value. Thus, in the ratio $(\bar{X} - \mu)/\sigma_{\bar{X}}$ there is only one statistic, $\bar{X}$, that varies from sample to sample. Since for normally distributed populations $\bar{X}$s are normally distributed, the z scores computed from this ratio are also normally distributed.

For large samples, $s_{\bar{X}}$ provides a close approximation of $\sigma_{\bar{X}}$, and may be taken as a substitute for $\sigma_{\bar{X}}$, and the distribution of values computed from the ratio $(\bar{X} - \mu)/s_{\bar{X}}$ closely resembles the normal distribution.

For small samples, there are two statistics in the ratio $(\bar{X} - \mu)/s_{\bar{X}}$ that vary, $\bar{X}$ and $s_{\bar{X}}$. Because the denominator $s_{\bar{X}}$ is not a fixed value but varies from sample to sample, the ratios are not exactly normally distributed; they are more widely dispersed than they would be in a normal distribution. This dispersion becomes more pronounced as the sample size becomes smaller.

Therefore, we cannot use the normal curve to represent the sampling distribution of $(\bar{X} - \mu)/s_{\bar{X}}$ for small samples when $s_{\bar{X}}$ is used as an estimate of $\sigma_{\bar{X}}$. The distributions that are appropriate in this situation are called t distributions. They were developed by William S. Gossett in the early part of this century. Gossett wrote under the pseudonym "Student"; the distributions he developed have been assigned the symbol t. Thus they are referred to as *Student's distributions of t* or, more simply, as *t distributions*.

The probabilities associated with differences between μ and $\bar{X}$, using $s_{\bar{X}}$, are derived from t distributions. The shape of the t distribution varies according to the number of degrees of freedom for a sample. Thus, instead of only one t distribution, there is a family of them, one for each sample based on its size. Figure 10-2 presents the t distributions for three different *df*s, and compares them to the normal curve. This figure shows that t distributions for large samples, such as those in which $df = 25$, more nearly correspond to the normal curve than do t distributions based on small samples, such as those in which $df = 9$

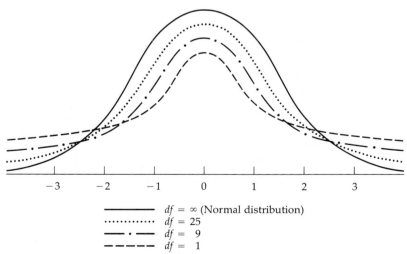

FIGURE 10-2. The t distributions for three different *df*s compared with the normal curve.

or $df = 1$. For small samples, the ratio $(\bar{X} - \mu)/s_{\bar{X}}$ is designated t rather than z, since z is reserved for use only with the normal curve.

The distributions of t and z are similar in that both have a mean of zero, are symmetrical, and are bell-shaped. They differ in that less of the t distribution's area is clustered close to the mean and more of its area is located in the tails.

What does all of this mean in terms of the area under the t-distribution curve? It means that, whereas we used z scores of -1.96 and 1.96 to specify the 95% confidence interval for the normal curve, we must go farther away from the mean toward the tails of a t-distribution curve to delineate the same area. To determine the t values associated with areas under the t-distribution curve, we first need to determine the df for the statistic. We have seen that the df is $N - 1$ for statistics based on the data in one sample.

Tables could be prepared giving the area under each t-distribution curve just as Table 1 was prepared for the normal curve. But this would be impractical, since we would need a separate table for each df. Instead, Table 2 in the back of the book gives the four most commonly used probability levels for each df. As was the case with z scores, t values can be positive or negative depending on whether they are above or below the mean. The values given in Table 2 represent both positive and negative t values. Since t-distribution curves are symmetrical, the area above a positive t is the same as the area below a negative t of the same value. Table 2 gives us the t values that cut off .10, .05, .02, and .01 of the area contained in *both* tails of each t distribution. This is sometimes called a table of two-tail values of t.

As an example of how to interpret Table 2, for $df = 21$, in the column headed $P = .05$, we find the t value of 2.080. This means that .05 of the area under the t-distribution curve based on 21 degrees of freedom lies outside of the area bounded by $t = -2.080$ and $t = 2.080$. Thus, to the left of $t = -2.080$ lies .025 of the area, and to the right of $t = 2.080$ lies .025 of the area. Between $t = -2.080$ and $t = 2.080$ lies .95 of the area under this particular curve.

Table 2 shows that t values are larger for small dfs than for large ones. We must go farther out on t-distribution curves based on small dfs than on the curves based on larger dfs to cut off the same area. The bottom row of this table gives t values for a sample with an infinite df. For $P = .05$, $t = 1.96$; this is exactly the value of z we used in determining the 95% confidence interval for a normally distributed sampling distribution. This illustrates the point made earlier, that the larger the sample size, the more nearly the t distribution follows the normal distribution. At infinity, they coincide.

Let's illustrate the use of the t distributions given in Table 2 to determine the 95% confidence interval for μ, based on data in a sample of ten individuals. Here $df = N - 1 = 9$. Since the 95% confidence interval cuts off $P = .05$ of the area, with this proportion equally divided between the two tails of the distribution, this is the value we use in computing the boundaries of the interval.

> **FORMULA 12**
>
> Calculation of confidence intervals.
>
> Lower Limit $\quad\bar{X} - ts_{\bar{X}}$
> Upper Limit $\quad\bar{X} + ts_{\bar{X}}$

Formula 12 gives the method for calculating the upper and lower limits of a confidence interval.

For $df = 9$ and $P = .05$, Table 2 indicates that $t = 2.262$. (Note that this is much larger than the $z = 1.96$ that we used with the normally distributed data.) Suppose that our sample has $\bar{X} = 30$ and $s_{\bar{X}} = 3$. The 95% confidence interval is determined by the values that lie at $-2.262 s_{\bar{X}}$ and $2.262 s_{\bar{X}}$.

The lower boundary lies at -2.262×3 from the mean of 30. This is $30 - 6.786 = 23.214$. The upper boundary lies at 2.262×3 from the mean. This is $30 + 6.786 = 36.786$. If we wish to determine the 99% confidence interval for the same data, Table 2 indicates that for $P = .01$, we should use a t value of 3.25. This will give us lower and upper boundaries of 20.25 and 39.75.

As is shown in Table 2, for $df = 29$, at $P = .05$, $t = 2.045$. This is not very different from $t = 1.96$ for $df = \infty$. However, this difference increases markedly as the dfs become smaller. To be perfectly accurate, we should always use the t distribution when we are dealing with sample means where $\sigma_{\bar{X}}$ is unknown, even though the t distributions for large dfs are almost identical with the normal distribution.

We will have more opportunity to use Table 2 as we explore other statistical techniques that involve sample means.

As evident above, the size of the three samples differ markedly, the means are identical and the standard deviations are virtually identical. But notice the difference in the value of the standard error of the means. The larger sample size has resulted in a smaller standard error statistic. Also, because the standard error is smaller for the larger sample, the effect on both the 95% and 99% confidence intervals is that they are narrower.

Figure 10-3 shows the 95% and 99% confidence intervals for the three samples, graphically illustrating the effect of sample size on the width of the intervals.

The computation of confidence intervals for these three samples of data shows that, for samples where the standard deviations are quite similar, there is an effect of sample size on the standard error of the mean and the width of the confidence intervals. Namely, larger samples yield smaller standard errors and narrower confidence intervals. Narrow confidence intervals give the researcher a smaller range within which to make an "educated guess" regarding the population mean.

To illustrate the effects of the sample size on the estimate of the standard error of the mean and, thus, on the width of the confidence interval, consider the following highly hypothetical situation.

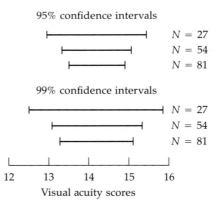

FIGURE 10-3. 95% and 99% confidence intervals for three sample sizes.

An educational researcher was interested in estimating the population mean reaction time of high school freshmen to visual cues provided by a popular video game. He obtained the reaction times "scores" on three randomly selected samples. The frequency distributions and statistics for these three samples are given below.

	Sample #1	Sample #2	Sample #3
20	1	2	3
19	1	2	3
18	2	4	6
17	2	4	6
16	3	6	9
15	4	8	12
14	4	8	12
13	2	4	6
12	3	6	9
11	1	2	3
10	2	4	6
9	1	2	3
8	1	2	3
N	27	54	81
Mean	14.185	14.185	14.185
S.D.	3.051	3.022	3.013
S.E.M.	.587	.411	.335
95% C.I.	12.958–15.412	13.346–15.024	13.502–14.868
99% C.I.	12.512–15.859	13.054–15.316	13.265–15.106

EXERCISES

1. A college professor randomly selected 30 sophomore women and gave each of them a social development self-test. She obtained the following test scores.

65	57	58	61	59	53
61	67	58	50	66	51
56	61	70	64	58	63
59	60	55	57	54	58
62	55	60	65	63	60

What is the 95% confidence interval for the population mean?

2. Using the data in Exercise 1, what is the 99% confidence interval for the population mean?
3. To estimate the mean length of time it takes fourth-grade children to read a short story, a researcher randomly selected 15 children and measured their reaction times. He calculated $\bar{X} = 18$ and $s = 1.3$. What is the 95% confidence interval for the population mean?
4. Using the data in Exercise 3, what is the 99% confidence interval for the population mean?
5. A manufacturer wanted to know about the variability in the weights of ball bearings produced by an automated production machine. He selected a random sample of ball bearings and obtained the following weights in grains.

13	24	23	14	23	17
20	12	15	20	17	20
20	27	19	14	19	23
17	19	13	26	16	19
22	17	22	16	23	19
17	25	16	23	17	20
19	21	18	21	19	20
16	21	15	24	20	24
10	11	10	18	11	18
21	18	24	12	21	21

What is the 95% confidence interval for the population mean?
6. Using the data in Exercise 5, what is the 99% confidence interval for the population mean?
7. A randomly selected group of 200 residents was asked to rate the quality of the schools in their city, giving a score of 100 for excellent and of 0 for poor. Their ratings yielded $\bar{X} = 75$ and $s = 9$. What is the 95% confidence interval for the population mean?
8. Using the data in Exercise 7, what is the 99% confidence interval for the population mean?

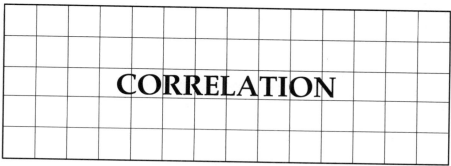

CORRELATION

As researchers, we are concerned with detecting relationships between and among phenomena. Many research studies are designed to find out whether there is an association between two variables. For example, we may want to determine if people's ages are related to their blood pressures, or whether students' anxiety levels are related to their achievement scores. By discovering a relationship between variables, we can often predict a person's status on a given variable if we know how he or she performs on the other variable. Since prediction is one of the major goals of any science, the discovery of relationships is of paramount importance. In this chapter, we present a way to show the relationship between two sets of data and one commonly used technique for measuring this relationship. In Chapter 12, we will demonstrate a method for making predictions about associated variables.

Note that the statistical techniques presented so far describe frequency distributions or make references from samples where scores were obtained for only one variable. In these cases, each datum represented a measurement based on only one characteristic, and we made statistical inferences using the sampling distribution that was appropriate for that statistic. Now we consider another very useful statistical technique that allows us to measure the relationship between two sets of data obtained from the same sample, or between data from two samples where individuals in the samples have been matched on some basis. For example, this technique permits us to specify the relationship between the pretest and posttest achievement test scores for fourth-grade students, or the motivation ratings and aptitude scores for a group of college freshmen, or the high school grade-point averages and the college senior grade-point averages for a group of students.

In such studies, we are not looking for differences between two groups of individuals; instead, we want to discover to what extent two sets of data are related. This statistical technique is called *correlation*. To illustrate its use, suppose we have the following arithmetic and spelling achievement scores for eight students:

Student	Arithmetic achievement scores	Spelling achievement scores
A	6	6
B	4	4
C	3	3
D	2	2
E	8	8
F	5	5
G	1	1
H	7	7

To depict graphically the correlation between the variables on arithmetic achievement and spelling achievement, we draw what is called a *scatter diagram*. To draw this diagram, which is shown in Figure 11-1, we choose one of the variables, for instance arithmetic achievement, to be represented on the vertical axis, and the other variable, spelling achievement, to be represented on the horizontal axis. Note that in this diagram the arithmetic scores are ranged along the vertical axis with the lowest score placed at the bottom, and the spelling scores are ranged along the horizontal axis with the lowest score placed at the left. This is the conventional method for arranging the scores in a scatter diagram, and it is consistent with the traditional Cartesian coordinate system.

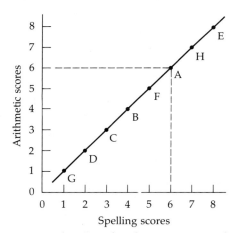

FIGURE 11-1. Scatter diagram of arithmetic scores and spelling scores.

Our data indicate that Student A received an arithmetic score of 6 and a spelling score of 6. To plot these two scores for Student A, we must locate both scores in the diagram and make one dot that will represent both scores. To do this, we locate value 6 on the arithmetic axis and extend a horizontal line from this position across the diagram. Next we locate value 6 on the spelling axis and extend a line vertically from this position. At the point where these two lines intersect we place a dot. This one dot now represents both scores for Student A, as Figure 11-1 shows.

Following the same procedure for each student in the group, we form a scatter diagram of the scores for all the students. We then find that we can draw a straight line connecting all of these dots in the scatter diagram in Figure 11-1.

This diagram shows that for every increase in score value on one variable, there is a corresponding increase on the other variable. Since this is true for every pair of scores in our data, we conclude that the relationship between arithmetic scores and spelling scores is *perfect*. In statistical terms, we would call this particular relationship a *perfect positive correlation*. It is called *perfect* because the amount of increase in a score on one variable is exactly proportional to the amount of increase in the score on the corresponding variable, with no exceptions. It is called *positive* because an *increase* in a score on one variable is associated with an increase in the score on the corresponding variable.

Now let's look at a scatter diagram that depicts a *perfect negative correlation*. Figure 11-2 is a scatter diagram that represents the relationship between the speeds of runners and the amounts of weight they are carrying. It is evident from this diagram that the speed of each runner is inversely related to the amount of weight he or she is carrying. Again in statistical terms, we would say that there is a perfect negative correlation between these two variables, because we can draw a straight line that runs through all of the dots in Figure 11-2. It is a *negative* correlation because the two variables are inversely related; that is, an increase in a score on one variable is associated with a decrease in the score on the corresponding variable.

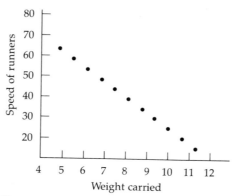

FIGURE 11-2. A perfect negative correlation.

To express the relationship between two variables statistically, we must have some numerical index showing the degree of correlation. This index is termed the *correlation coefficient*, and its magnitude indicates the degree to which two frequency distributions of data are related.

Many different correlation techniques are available to the statistician. Which one is appropriate for use in a particular situation depends on the nature of the data being analyzed. This chapter presents a method for computing one type of correlation coefficient—the *Pearson product-moment correlation coefficient*. It is named after its originator, Karl Pearson, and is derived by examining functions of deviations of values from the "best-fitting" line. The term *moment* is taken from the science of mechanics and refers to certain functions of deviations. The symbol for this correlation coefficient is r. It is one of the more commonly used correlational techniques. To use it properly, however, we must assume that the variables are linearly related, and that the scores on each variable come from normally distributed populations. If these assumptions cannot be made, this type of correlation analysis is inappropriate and other techniques must be used. (In Chapter 19 we examine a correlation technique that can be used for data that do not meet these requirements.)

The coefficient of correlation for the perfect positive correlation shown in Figure 11-3(a) is $r = 1.00$. The coefficient for the perfect negative correlation shown in Figure 11-3(b) is $r = -1.00$. These are the maximum values for r.

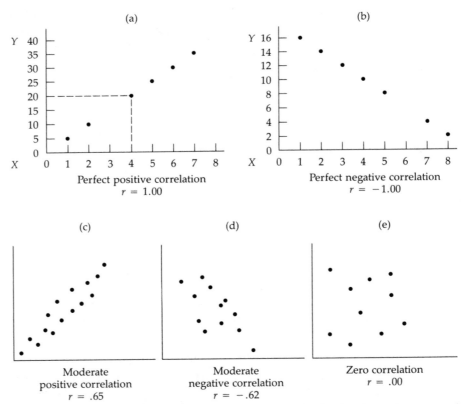

FIGURE 11-3. Scatter diagrams depicting various levels of correlation.

Note that the sign of the correlation coefficient indicates whether the correlation is positive or negative and that the size of the perfect correlation is the same (1.00) regardless of whether it is positive or negative. This is important to remember because a common mistake is to think that a coefficient of $r = -1.00$ represents no correlation. The coefficient that indicates no degree of correlation is $r = .00$. This condition occurs when scores on one variable are not related in any way to scores on the other variable. Figure 11-3(e) is a scatter diagram of uncorrelated data, representing the relationships between the IQs of soldiers and their rifle marksmanship scores.

As you may imagine, a perfect correlation between two variables rarely occurs. Almost every time a relationship exists between two variables, it is less than perfect. In such cases, the coefficient is less than 1.00. For example, $r = .85$ indicates that there is a fairly strong positive correlation between two variables, $r = .54$ indicates that the positive correlation is not as strong, and $r = .03$ indicates that there is practically no positive correlation. Likewise, $r = -.75$ indicates a fairly strong negative correlation, and $r = -.12$ indicates a weak negative correlation. Thus we see that all positive coefficients indicate direct relationships and all negative coefficients indicate inverse relationships, and that the size of the coefficient indicates the strength of the relationship. Figures 11-3(c) and 11-3(d) depict scatter diagrams showing moderate positive and negative correlations.

Suppose we obtain IQs and reading scores for a group of students and prepare the scatter diagram shown in Figure 11-4. We can see that the dots on the diagram tend to lie in a positive direction, although they certainly do not lie in a straight line. This indicates that the correlation is positive, but less than perfect. We can compute the correlation coefficient for these data; in this case, $r = .75$.

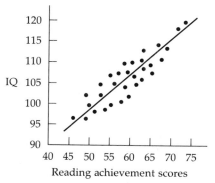

FIGURE 11-4. Scatter diagram of IQs and reading scores.

Let's see how we determine the size of a correlation from the scatter of the dots in the diagram. First we draw a straight line through the dots that best represents the linear trend shown in the diagram. This line is positioned in the scatter diagram so that the average distance of the dots from it is as small as possible, as Figure 11-4 shows. If we measure the perpendicular distance of each dot from this line, square each distance, and sum these squared distances, the sum will be smaller than the sum we could obtain by placing the line in any other position in the diagram. A line placed in this fashion is called a *best-fitting* line.

The total of the distances that the dots lie from this best-fitting line is inversely related to the size of the correlation coefficient. For example, if the dots are widely scattered, the distances of the dots from the best-fitting line are great and the size of the coefficient is small. On the other hand, if the dots deviate very little from the best-fitting line, the coefficient is large. If there is no deviation from the best-fitting line, as in Figures 11-3(a) and 11-3(b), the coefficient is either 1.00 or −1.00.

COMPUTATION OF THE CORRELATION COEFFICIENT

We do not need to prepare a scatter diagram to determine the degree of correlation between two variables; we have a statistical procedure that allows us to compute the correlation coefficient directly from the data. However, a scatter diagram is helpful because it gives us a visual indication of the linearity of the relationship and the variability of the data on each variable.

Formula 14 presents the formula for calculating the Pearson product-moment correlation coefficient.

FORMULA 14

Calculation of the Pearson product-moment correlation coefficient.

$$r = \frac{N\Sigma XY - (\Sigma X)(\Sigma Y)}{\sqrt{[N\Sigma X^2 - (\Sigma X)^2][N\Sigma Y^2 - (\Sigma Y)^2]}}$$

N = number of pairs of scores
$df = N - 2$

Since we are dealing with two sets of data, we generally assign the symbol X to the scores on one variable and the symbol Y to the scores on the other variable. The expression $N\Sigma XY$ in the numerator of Formula 14 indicates that we obtain the product of each pair of scores, sum these products, and then multiply this sum by the number of pairs of scores (N).

To illustrate the application of Formula 14 with a simple example, suppose a teacher wishes to determine if the scores fourth-grade students obtain on a spelling test are directly related to their reading scores. The teacher obtains spelling scores (X) and reading scores (Y) for 12 fourth-grade students; these scores are shown in Table 11-1. The computation of the correlation coefficient using Formula 14 is given below the table.

TABLE 11-1. Computation of the Pearson product-moment correlation coefficient

Spelling scores X	Reading scores Y	X^2	Y^2	XY
1	1	1	1	1
2	4	4	16	8
3	2	9	4	6
4	4	16	16	16
4	6	16	36	24
5	2	25	4	10
6	3	36	9	18
6	7	36	49	42
7	5	49	25	35
8	4	64	16	32
8	9	64	81	72
9	8	81	64	72
$\Sigma X = 63$	$\Sigma Y = 55$	$\Sigma X^2 = 401$	$\Sigma Y^2 = 321$	$\Sigma XY = 336$

Using Formula 14: $r = \dfrac{12(336) - (63)(55)}{\sqrt{[12(401) - (63)^2][12(321) - (55)^2]}} = .679$

CONCLUSION

The magnitude of a correlation coefficient can be affected by several factors. First, if the relationship between the variables is not linear, the Pearson product-moment correlation coefficient yields an underestimate of the true relationship between the variables. There are other statistical methods for determining the correlation between variables that have curvilinear relationships. Also, the correlation between variables with large variances tends to be greater than the correlation between variables that have a curtailed range of values.

A word of caution is needed on the interpretation of a correlation coefficient. Although the coefficient indicates the degree to which two variables are related, this does not necessarily mean that there is a causal relationship between them. Correlation does not imply that one variable is causing the variation in the other. A simple example illustrates that correlation cannot be interpreted in this way. Suppose we find a correlation between children's neatness of appearance and their punctuality in arriving at school. By no stretch of the imagination can we say that being neat causes the children to be on time or that being punctual causes them to be neat. This relationship may actually be caused by a third variable, such as the kind of parental attention the children receive.

Another caution should also be observed in interpreting correlation coefficients. Because they are indexes of relationship, they cannot be interpreted as percentages of agreement. A coefficient of $r = .30$ does not indicate that there is a 30% agreement between the two sets of scores. Also, it is not proper to say that a correlation of $r = .40$ is twice as strong as a correlation of $r = .20$ just because it is twice as large.

If we want to make statements regarding the meaning of a correlation, we can compute the *coefficient of determination*, which is obtained by squaring the correlation coefficient. This value, r^2, can be properly interpreted as the proportion (or percentage) of the variance of the X scores that is associated with the variance of the Y scores. To illustrate, in the example in Table 11-1, where $r = .679$, the coefficient of determination $r^2 = (.679)^2 = .461$ indicates that 46% of the variance of the X scores is accounted for by the variance of the Y scores.

EXERCISES

1. An industrial arts teacher who wanted to determine if there was any relationship between students' mechanical comprehension and their divergent thinking ability randomly selected a group of students and measured them on these two variables.

Mech.	Diver.	Mech.	Diver.	Mech.	Diver.
15	35	27	44	19	37
20	39	21	37	21	41
25	41	11	31	15	38
16	33	26	41	25	42
12	32	17	36	14	32
26	44	19	35	23	39
22	38	22	40	16	34
14	34	13	34	22	43
24	40	24	42	13	31
18	37	15	36	19	41

   a) Compute the Pearson product-moment correlation coefficient.
   b) Compute the coefficient of determination. What does it indicate about the relationship between the variables of mechanical comprehension and divergent thinking in this sample?
   c) Prepare a scatter diagram of these data.

2. An elementary school teacher wanted to see if there was a relationship between how long it took her pupils to complete a spelling quiz and the accuracy of their answers. She obtained the following data from the pupils in her class.

Spelling scores	Time (minutes)	Spelling scores	Time (minutes)
25	5.5	31	4.2
26	5.4	33	3.6
28	5.7	35	5.0
28	4.6	37	4.7
29	4.9	42	4.0
30	4.8	43	4.9
30	4.9		

CORRELATION

a) Compute the Pearson product-moment correlation coefficient.
b) Compute the coefficient of determination. What does it indicate about the relationship between spelling ability and completion times?
c) Prepare a scatter diagram of the relationship between these two variables for this sample.

REGRESSION

We have seen that the correlation coefficient is a useful statistical index for describing the degree of relationship between two variables. It follows that if two variables are correlated, we should be able to estimate the score an individual would obtain on one variable if we know his or her score on the related variable. For example, if we know that social competence is positively correlated with chronological age, we can predict that 8-year-old children will receive higher social competence scores than 5-year-old children. However, this estimate usually is not precise enough for our purposes. We need a method for predicting the values of social competence scores for children at various age levels. For the data given in Figure 11-1, we could "predict" that an individual with a spelling score of 5 would obtain an arithmetic score of 5. This prediction could be made with complete accuracy, because the correlation between these two variables was perfect. Variables in the behavioral sciences seldom correlate perfectly, but even with less than perfect correlations, we can make a fairly good prediction of an individual's score on one variable, if we are given his or her score on another related variable, by a statistical method known as *regression analysis*.

Through the formulas used in regression analysis, we can make theoretical predictions based on examination of the relationship between the data on the predictor variable and the data on the predicted variable from a sample of individuals. In this context, the term *prediction* is not limited solely to making future projections; it also means estimating a person's present status on one variable, given his present status on the related variable.

Let's first consider the situation in which two variables are linearly related. The method we will use is called *linear regression*; it is based on our ability to place a best-fitting straight line, called a *regression line*, through two sets of correlated data. We use a *regression equation* to describe the regression line mathematically.

THE REGRESSION OF Y ON X

In this regression analysis, we assign the symbol X to the *predictor* variable, and Y to the *predicted* variable. In research terminology, the predictor is called the *independent variable* and the predicted variable is called the *dependent variable*. The regression equation permits us to predict an individual's Y score if we know his X score. To illustrate this procedure, we can use the data presented in Example 12-1.

EXAMPLE 12-1 A school librarian noticed that students who spend a good deal of time in the library tend to receive higher achievement scores than those who use the library infrequently. He wished to establish a method for predicting a student's achievement score once he knew the number of hours the student spent in the library. The librarian collected the following data on 24 students:

Average number of hours per week in library (X)	Achievement test scores (Y)	Average number of hours per week in library (X)	Achievement test scores (Y)
1	1	6	3
2	1	6	5
2	3	6	7
2	4	7	5
3	2	7	5
3	2	7	7
4	4	8	4
4	4	8	8
4	6	8	9
5	2	9	7
5	3	9	8
5	6	9	9

• • • • • • • •

In Example 12-1, the predictor variable (X) is the average number of hours per week students spend in the library (independent variable) and the predicted variable (Y) is their achievement scores (dependent variable). Figure 12-1 shows the scatter diagram for these data.

Suppose we want to predict the achievement score for a student who spends an average of two hours per week in the library. The data in this study indicate that three students in the sample averaged two hours per week in the library, and they received achievement scores of 1, 3, and 4. Therefore, one way to predict this student's score is to take the average of these three scores, which is 2.67, and use it as the predicted score. By this process, we predict that a student averaging two hours per week in the library will receive an achievement score of 2.67. But this estimate is based on a sample of only three students! Logic tells us that a prediction based on such a small sample is of

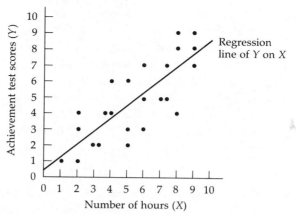

FIGURE 12-1. Scatter diagram of achievement test scores and number of hours in library, with the regression line of Y on X.

limited value because of the amount of sampling error that is likely to be involved. We would feel more confident if we were to use all the data obtained in our sample to determine the predicted value. We need to establish the trend of the Y scores as they relate to the X scores, and this trend should be derived from all the available data.

Our first task, then, is to determine the best-fitting line in this scatter diagram that can be used to predict Y scores on the basis of X scores. Such a line is called the *regression line of Y on X*. This line should be positioned in the diagram so that the sum of squares of the vertical distances of the data from the regression line is as small as possible. (This regression line is shown in Figure 12-1.) To position this line precisely in the scatter diagram, we must locate at least two points on it. The regression line can then be drawn through these two points. To locate points on the regression line, we use the regression equation, which is given in Formula 16.

FORMULA 16

Calculation of the regression equation.
$$\tilde{Y} = a + b_{yx}X$$
where: $\tilde{Y}$ = the predicted value of Y

In Formula 16, the symbol $\tilde{Y}$ indicates a predicted value of Y that corresponds to a given X. The symbol b_{yx} is called the *regression coefficient of Y on X*, and the symbol a is called the *Y intercept*. First, we will illustrate how this formula is used to position the regression line in the scatter diagram, and then we will present the formulas used to compute b_{yx} and a.

The regression coefficient b_{yx} describes the *slope of the best-fitting line* by specifying the amount of increase on the Y variable that accompanies one unit of increase on the X variable. The regression coefficient can be described as b_{yx} = vertical change/horizontal change.

Figure 12-2 shows a regression line of Y on X developed from data on two related variables. The actual scatter of the data around this regression line has been omitted in the figure. The regression of Y on X in Figure 12-2 indicates that for every 10 units of horizontal change (on the X variable), there are 5 units of vertical change (on the Y variable). Therefore,

$$b_{yx} = 5 \text{ units}/10 \text{ units} = .5,$$

which is the ratio of vertical change to horizontal change.

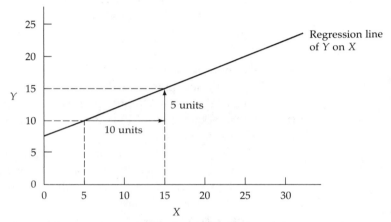

FIGURE 12-2. Regression line of Y on X (data omitted).

The b_{yx} indicates the amount of increase in Y that is accompanied by an increase in X. A negative b_{yx} value indicates that an increase in Y is accompanied by a *decrease* in X.

In Formula 16, a is called the Y *intercept*, and it represents the value of Y that corresponds to X = 0; that is, it is the value of Y at the point where the regression line crosses the Y axis. In Figure 12-2, the regression line of Y on X intersects the Y axis at Y = 7; that is, when X = 0, Y = 7. Therefore, in this example, $a = 7$.

From the regression line in Figure 12-2 we have determined the values of b_{yx} and a. By substituting these values in Formula 16, we can state the regression equation for this figure as $\tilde{Y} = 7 + .5X$. This equation can be used to obtain the predicted value of Y (symbolized by $\tilde{Y}$) for a given X value by substituting the X value in the equation and solving for $\tilde{Y}$. For example, for X = 25, $\tilde{Y} = 7 + .5(25) = 19.5$. Here we have used the regression equation to predict that a person receiving an X of 25 will have a Y of 19.5.

We have used Figure 12-2 to illustrate how we determine b_{yx} and a by examining the placement and slope of the regression line in a scatter diagram. In actual practice, however, to draw the regression line for Y on X, we proceed in reverse fashion, using the obtained data on two variables to determine the values of b_{yx} and a, which we then use to form the regression equation. Thus to develop the regression equation from the data given in Example 12-1 and to draw the regression line in the scatter diagram in Figure 12-1, we need to compute the values represented by b_{yx} and a.

Formula 17 shows the calculation of the regression coefficient of Y on X from sample data.

FORMULA 17

Calculation of the regression coefficient of Y on X.

$$b_{yx} = \frac{\Sigma XY - \frac{(\Sigma X)(\Sigma Y)}{N}}{\Sigma X^2 - \frac{(\Sigma X)^2}{N}}$$

Applying this formula to the data in Example 12-1, where $\Sigma X = 130$, $\Sigma Y = 115$, $\Sigma X^2 = 844$, $\Sigma Y^2 = 689$, and $\Sigma XY = 733$, we obtain:

Using Formula 17: $\quad b_{yx} = \dfrac{733 - \dfrac{(130)(115)}{24}}{844 - \dfrac{(130)^2}{24}} = .79$

Formula 18 is the formula for calculating the value of the Y intercept.

FORMULA 18

Calculation of the Y intercept.

$$a = \bar{Y} - b_{yx}\bar{X}$$

In Example 12-1, $\bar{Y} = 4.79$ and $\bar{X} = 5.42$. We determine the value of the Y intercept as follows:

Using Formula 18: $\quad a = 4.79 - .79(5.42) = .51$

Having calculated both b_{yx} and a, we can use Formula 16 to state the regression equation of Y and X for Example 12-1.

Using Formula 16: $\quad \tilde{Y} = .51 + .79X$

Now we can use this regression equation to predict the value of Y for any selected value of X. Using the data in Example 12-1, suppose we want to predict the achievement score for a student who has used the library for two hours per week, or for eight hours per week. Using the regression equation:

$$\text{For } X = 2: \quad \tilde{Y} = .51 + .79(2) = 2.09$$
$$\text{For } X = 8: \quad \tilde{Y} = .51 + .79(8) = 6.83$$

To draw the regression line of Y on X in Figure 12-1, we simply place a dot at $X = 2$, $Y = 2.09$ and a dot at $X = 8$, $Y = 6.83$, and we draw the regression line through these two dots.

Another way of locating the regression line is based on the fact that the line always passes through the point in the scatter diagram where the means of the two variables coincide. Therefore, we know that the regression line passes through the point at which $\bar{X} = 5.42$ and $\bar{Y} = 4.79$. We learned earlier that a gives the value of Y corresponding to $X = 0$. In our example, $a = .51$, so we know that the regression line also passes through the point at which $X = 0$ and $Y = .51$. Therefore, using the mean values and a, we can determine two points in the scatter diagram through which we can draw the regression line. The regression line can be used directly to obtain a predicted Y value for any given X value merely by locating the point on the line directly above the X value and reading the corresponding Y value on the vertical axis.

It should be obvious that we do not have to actually prepare a scatter diagram and draw the regression line to be able to predict Y scores. All we have to do is to use Formulas 16, 17, and 18 to develop the regression equation, and then use this equation to calculate $\tilde{Y}$ for any X value.

THE REGRESSION OF X ON Y

Up to this point we have been concerned with defining one regression line, that of Y on X. It is possible to develop a second regression equation that will permit us to predict X values for given values of Y. This is appropriate only when designating the variables as dependent and independent is interchangeable. For example, if we find that speed of typing and speed of taking shorthand dictation are correlated, we could develop a regression equation to predict an individual's typing speed if we know the shorthand speed, or we could develop a regression equation to predict shorthand speed if we know the typing speed.

The second regression line and its regression equation describe the regression of X on Y. The line depicting the regression of X on Y differs from the regression line of Y on X whenever the correlation between the two variables is less than perfect. Only when $r = 1.00$ or -1.00 will the two regression lines be identical.

The formulas for predicting X values corresponding to given Y variables are Formulas 19, 20, and 21.

Formulas 19 through 21. Formulas for calculation of the regression of X on Y.

Formula 19 $\quad \tilde{X} = a + b_{xy}Y \quad$ (regression equation)

Formula 20 $b_{xy} = \dfrac{\Sigma XY - \dfrac{(\Sigma X)(\Sigma Y)}{N}}{\Sigma Y^2 - \dfrac{(\Sigma Y)^2}{N}} \quad$ (regression coefficient)

Formula 21 $\quad a = \bar{X} - b_{xy}\bar{Y}$

where: $\tilde{X}$ = the predicted value of X.

Formula 19 gives the regression equation for X on Y, with the predicted value of X symbolized by $\tilde{X}$. Formula 20 is the formula for determining the regression coefficient of X on Y, which is symbolized by b_{xy} (rather than b_{yx}, as in Formula 17). The numerator of this formula is identical to the numerator of Formula 17, but the denominator represents the sum of squares for the Y variable. Formula 21 shows how to calculate the X intercept, and it yields the value of X corresponding to Y = 0. Thus it gives the value of X at the point where the regression line crosses the Y axis. The regression lines cross each other at the point in the scatter diagram where $\bar{X}$ and $\bar{Y}$ coincide.

To illustrate the computation of the regression equation for X on Y and the location of the regression line, suppose the librarian in Example 12-1 wanted to predict the average number of hours a student used the library corresponding to a given achievement score. (This seems somewhat unlikely, but such a prediction can be made using regression techniques.) Applying Formulas 19, 20, and 21 to the data in Example 12-1, we obtain:

Using Formula 20: $\quad b_{xy} = \dfrac{733 - \dfrac{(130)(115)}{24}}{689 - \dfrac{(115)^2}{24}} = .80$

Using Formula 21: $\quad a = 5.42 - .80(4.79) = 1.59$

Using Formula 19: $\quad \tilde{X} = 1.59 + .80Y$

We can use the regression equation given in Formula 19 to predict X values corresponding to given Y values. For example, we can predict that a student who received an achievement score of Y = 4 spent an average of $\tilde{X} = 1.59 + .80(4) = 4.79$ hours per week in the library. We can predict that the average weekly library time for a student with Y = 9 to be $\tilde{X} = 1.59 + .80(9) = 8.79$ hours.

Figure 12-3 depicts the same scatter diagram as Figure 12-1, but it shows both regression lines. Note that the regression line of X on Y crosses the X

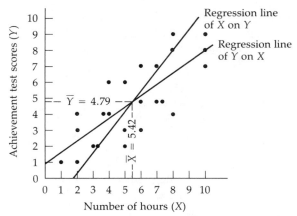

FIGURE 12-3. Scatter diagram of achievement test scores and number of hours in library, with regression lines of Y on X and X on Y.

axis at $X = 1.59$ (the value of a, which is computed by using Formula 21), and also that the two regression lines intersect at the point where the means of the X and Y scores coincide.

Although we have shown that regression equations can be used to predict values on both the X and Y variables, we must stress that it would be foolish to conclude that a student with a particular X value would, in fact, have the precise value of Y that the equation predicted. A more correct interpretation is that the average Y score of students who have a given X score will tend to be close to the predicted Y value. Thus the regression line can be thought to represent a kind of continuous mean that gives us the expected value, or mean, of Y for a particular X value. In fact, there are statistical techniques that permit us to develop confidence intervals for the population mean of Y for any given value of X (and for the population mean of X for any given value of Y).

Recall the example in Chapter 11 in which we determined the correlation between hostility and aggression scores for a group of adolescent boys. We can also develop a regression equation that will permit us to predict a boy's aggression score if we know his hostility score (or vice versa). The regression equation for these data is

$$\tilde{Y} = -1.319 + .572X$$

Using this equation, you can calculate the predicted Y scores (aggression) for any given value of X (hostility).

To actually draw the regression line, first prepare a scatter diagram of the data, and then locate the point in the diagram representing the mean scores of the two variables. The mean hostility score is 45.750 and the mean aggression score is 24.850. The line will pass through this point.

To draw the regression line, you must locate one other point in the diagram. Suppose we predict the Y score for $X = 50$, which is the predicted $Y = 27.281$. Locate the point in the diagram representing $X = 50$ and $Y = 27.281$. Now you have two points through which to draw the regression line. Your scatter diagram and regression line should look like those in Figure 12-4.

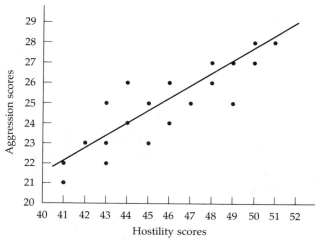

FIGURE 12-4. Scatter diagram of hostility scores and aggression scores and regression line.

MULTIPLE REGRESSION

In the previous section, we examined methods for predicting a value on one variable when we specify a value on another related variable. Regression analysis can be extended to include two or more predictor variables. It is logical to assume that, if two predictor variables are correlated with the criterion variable, using both of them in a regression equation will improve the accuracy of prediction. Indeed, the statistical technique of multiple regression allows us to combine two or more predictor variables in establishing a regression equation. We will introduce the two-predictor situation here and provide the computational formulas for developing the multiple regression equation. Consider the situation in Example 12-2.

EXAMPLE 12-2 Two personel supervisors interview a group of salespersons to determine the supervisors' accuracy in predicting which salespersons will exhibit greater sales volume during a given year. Each supervisor, here designated X_1 and X_2, provides a potential sales rating for each employee. At the end of a given year, each salesperson's actual sales record, designated as Y, is obtained. The following data were obtained.

First supervisor's rating (X_1)	Second supervisor's rating (X_2)	Sales record (Y)
10	20	62
15	27	59
17	25	63
18	17	60
20	12	58
22	33	70
23	25	73
23	27	69
25	28	75
27	25	80
29	34	74
32	29	77
34	20	79

• • • • • • • •

In multiple regression analysis, we establish a regression equation which allows us to use two predictors X_1 and X_2 in predicting the criterion Y. The form of this equation is an extension of the form used in the one-predictor situation and is presented as Formula 22.

FORMULA 22

> Calculation of the multiple regression equation (two predictors).
>
> $$\tilde{Y} = a + b_1 X_1 + b_2 X_2$$
>
> where X_1 and X_2 are the two predictor variables.

To develop the equation in Formula 22, we must calculate the value of a (the regression constant) and the values of b_1 and b_2 (called the partial regression coefficients). Then, to use the regression equation for predicting a value of Y, we must enter values for X_1 and X_2 and solve for Y.

The formulas for calculating a, b_1, and b_2 involve the means and standard deviations of each of the three variables ($\bar{Y}$, $\bar{X}_1$, and $\bar{X}_2$) and the correlations between each pair of variables. The means and standard deviations for these data are:

For first supervisor (X_1)	$\bar{X}_1 = 22.692$	$s_1 = 6.836$
For second supervisor (X_2)	$\bar{X}_2 = 24.769$	$s_2 = 6.193$
Salesperson's sales (Y)	$\bar{Y} = 69.154$	$s_Y = 7.904$

The correlations are:

First supervisor vs. second supervisor	$r_{12} = .315$
First supervisor vs. sales record	$r_{1Y} = .852$
Second supervisor vs. sales record	$r_{2Y} = .483$

The subscripts on the symbols are used to denote the variable or variables. Note that the two supervisors' ratings were moderately correlated (.315). Also, the first supervisor's ratings were much more correlated with the sales records than were the second supervisor's ratings. This indicates that the ratings of the first supervisor should be given more "weight" in predicting sales than those of the second supervisor.

The formulas for calculating a, b_1, and b_2 are given as Formulas 23, 24, and 25.

FORMULAS 23 through 25

Formulas for developing the multiple regression equation.

Formula 23
$$b_1 = \left(\frac{r_{1Y} - (r_{2Y})(r_{12})}{1 - r_{12}} \right) \left(\frac{s_Y}{s_1} \right)$$

Formula 24
$$b_2 = \left(\frac{r_{2Y} - (r_{1Y})(r_{12})}{1 - r_{12}} \right) \left(\frac{s_Y}{s_2} \right)$$

Formula 25
$$a = \bar{Y} - [(b_1)(\bar{X}_1) + (b_2)(\bar{X}_2)]$$

For the data in Example 12-2, we obtain the following:

$$b_1 = \left(\frac{.852 - (.483)(.315)}{1 - .315} \right) \left(\frac{7.904}{6.836} \right) = .898$$

$$b_2 = \left(\frac{.483 - (.852)(.315)}{1 - .315} \right) \left(\frac{7.904}{6.193} \right) = .304$$

$$a = 69.154 - [(.898)(22.692) + (.304)(24.769)] = 41.237$$

Thus the multiple regression equation is obtained by substituting the preceding values in Formula 22.

$$\tilde{Y} = 41.237 + .898X_1 + .304X_2$$

Examination of the partial regression coefficients b_1 and b_2 indicate that b_1 is the larger of the two. This was expected because the first supervisor ratings were more highly correlated with sales. Thus the X_1 scores, when entered into the regression equation, will have more of an effect on the predicted Y scores than will the X_2 scores.

This regression equation may be used to predict a value for Y for any given values of X_1 and X_2. For example, to predict the sales record for a salesperson having a rating of 21 by the first supervisor and 30 by the second supervisor you need to substitute these values in the equation and solve for Y:

$$Y = 4.237 + .898(21) + .304(30)$$
$$Y = 4.237 + 18.858 + 9.120$$
$$Y = 32.215$$

Thus, a sales record of 32.215 would be predicted for an individual receiving these two supervisor ratings.

EXERCISES

1. A random sample of 14-year-old educable mentally retarded children was selected, and each child's mother and father were asked to rate the child's problem-solving ability. The following ratings were obtained.

Mother rating	Father rating	Mother rating	Father rating
5	10	25	35
10	15	30	25
10	20	35	35
15	5	40	25
15	15	40	35
20	15	40	45
20	25	45	35
25	20	50	25

   Using X for the mother ratings and Y for the father ratings, do the following:
   a) Determine the regression equation for father ratings on mother ratings.
   b) Using the regression equation, predict the father rating for an educable mentally retarded child who receives a mother rating of 15. For one who receives a mother rating of 40.
   c) Construct a scatter diagram of the preceding data and draw in the regression line of Y on X.
   d) Determine the regression equation of mother ratings on father ratings.
   e) Using the regression equation, predict the mother rating for a child who receives a father rating of 20. For one who receives a father rating of 45.
   f) In the scatter diagram, draw the regression line of X on Y.

2. A university admissions officer wanted to predict the university graduates' final grade point averages (U-GPA) by using their high school grade point averages (HS-GPA) and their university entrance examination scores (U-EE) as predictors. The admissions officer collected the following data from a random sample of recent university graduates.

U-GPA	HS-GPA	U-EE
2.5	2.8	22
2.4	2.8	23
2.3	2.9	23
2.5	2.9	25
2.7	3.0	24
2.8	3.0	26
3.5	3.1	23
3.2	3.1	26
2.9	3.2	24
2.9	3.2	27
3.0	3.3	23
3.1	3.3	26
3.5	3.4	25
3.7	3.4	27
2.6	3.5	25
3.5	3.5	28
2.8	3.6	27
3.9	3.6	29

a) Determine the regression equation for predicting U-GPA scores.
b) Predict the U-GPA of a graduate who had a HS-GPA of 2.4 and a U-EE of 23.
c) Predict the U-GPA of a graduate who had a HS-GPA of 3.4 and a U-EE of 29.

3. Exercise 1 of Chapter 11 presented the mechanical comprehension and divergent thinking scores of a group of students.
   a) Develop a regression equation to predict the divergent thinking scores for selected mechanical comprehension scores.
   b) What is the predicted divergent thinking score for a student who has a mechanical comprehension score of 14? What is predicted for a score of 24?
   c) Develop a regression equation to predict the mechanical comprehension scores for selected divergent thinking scores.
   d) What is the predicted mechanical comprehension score for a student who has a divergent thinking score of 32? What is predicted for a score of 42?
   e) Prepare a scatter diagram (or use the one you prepared earlier) and plot the regression line of divergent thinking scores on mechanical comprehension scores.
   f) In the scatter diagram, plot the regression line of mechanical comprehension scores on divergent thinking scores.

4. A psychology student was given the assignment to develop a method for predicting the speed with which white rats would complete a maze based on the amount of drug LM-22 injected and the illumination level of the maze. She selected a group of rats and randomly injected varying amounts of the drug and randomly changed the illumination level as she ran them through the maze. She obtained the following statistics.

Time to complete maze	Mean = 62.613	s.d. = 10.443
Drug LM-22 injected	Mean = 17.621	s.d. = 4.310
Amount of illumination	Mean = 32.116	s.d. = 6.919

   The correlations were:

Drug vs. time	$r = .593$
Illumination vs. time	$r = -.417$
Drug vs. illumination	$r = .174$

   a) What is the regression for predicting time to run the maze for given values of drug dosage and amount of illumination?
   b) What is the predicted time for a rat given a dosage of 15 and an illumination of 27? For one with a dosage of 22 and illumination of 35?

13 INTRODUCTION TO HYPOTHESIS TESTING

The statistician engages in two forms of statistical inference—estimating population parameters and testing hypotheses. In forming confidence intervals, we were involved in estimating the parameter μ from sample data. In this chapter, we introduce the other major function of statistics, that of providing a basis on which to make a decision about the tenability of a hypothesis.

In practical research, where we are constantly testing hypotheses, we usually want to compare two or more samples. For example, we may want to determine whether there is a difference in achievement between students who are taught by the lecture method and those who are taught by the discussion method, or whether boys can jump farther than girls. Sometimes we may want to make comparisons among more than two groups; but first, let's consider the situation in which only two groups are to be compared.

Suppose we randomly select two samples of fourth-grade children, with 50 children in each sample, and teach them music appreciation by two different methods. For convenience, we designate these Method A and Method B. At the end of the school year, we administer a music appreciation test to both groups and obtain scores for both samples. We compute the mean and standard deviation for both samples and obtain the following data:

	Method A	Method B
N	50	50
$\bar{X}$	75	79
s	7	8

From these data, we can tell that the group of students taught by Method B (we will call this group Sample B) received a higher mean score than the group taught by Method A (called Sample A). If we knew that these sample means were identical to the population means for the two methods, we could say that Method B was superior to Method A. But from our previous discussion, we know that sampling error is always involved when we select a sample from a population. (In this case, the population is all fourth-grade children who might be taught music appreciation by Method A or Method B.) Even if we take two random samples from the same population, the means will be different, because of sampling error. Therefore we need to know by how much the means must differ before we can assume that they are derived from different populations. In other words, the question that we, as statisticians, ask is: "What is the probability that the difference between the two sample means is due to sampling error?" Can the difference between the sample means for Sample A and Sample B be attributed to random error in our sampling, or do children taught by one method actually learn more than those taught by the other method? In effect, are we dealing with two different populations?

Before we examine the procedure used to answer this question, we must consider how a researcher states a research hypothesis and how a statistician deals with it. In our example, the research question to be answered is:

Research question: Is there a difference in effectiveness between Method A and Method B for teaching music appreciation to fourth-grade students?

Restated as a research hypothesis, this question becomes:

Research hypothesis: There is a difference between the effectiveness of Method A and the effectiveness of Method B for teaching music appreciation to fourth-grade students.

To make a statistical analysis of the data, we must state the research hypothesis in statistical terms. The hypothesis might then read:

Research hypothesis: The mean music appreciation score for the population of fourth-grade students taught by Method A is different from the mean score of those taught by Method B.

We run into an interesting problem when we attempt to test the validity of this hypothesis by applying statistical techniques to data obtained from two samples given different methods. The problem is that the research hypothesis is nonspecific in that it does not state the size of the difference in effectiveness. We need a hypothesis that is explicit enough to be testable. The research hypothesis can be restated to make it specific by converting it into a *null hypothesis*. It might then read:

Null hypothesis: There is no difference in effectiveness between Method A and Method B for teaching music appreciation to fourth-grade students.

To a statistician, this is saying that the population from which Group A was selected has the same mean as the population from which Group B was selected, and that the obtained difference between the two sample means is due to nothing more than sampling error.

The preceding null hypothesis can be expressed symbolically as:

$$H_0: \quad \mu_1 - \mu_2 = 0$$

This is read: "The null hypothesis is that the difference between the mean of population 1 and the mean of population 2 equals zero."

The research hypothesis can be expressed as:

$$H_1: \quad \mu_1 - \mu_2 \neq 0$$

This is interpreted as: "The research hypothesis is that the difference between the means of the two populations is not equal to zero."

Although most null hypotheses state that the difference between population means is zero, other kinds of statements can also be tested. A null hypothesis can specify, for example, that a difference between population means is 10 points. In this case, the research hypothesis might be that the difference between population means is not 10 points. In this text, we will be testing only the null hypothesis of no difference, but you should be aware that the statement of a null hypothesis can also take other forms.

Returning to our example, we have made the null hypothesis of no difference in effectiveness between Method A and Method B. Our task is to decide whether or not to reject the null hypothesis. To help us decide, we apply statistical techniques to the data in the two samples; this enables us to determine the probability that the null hypothesis is false. We decide not to reject the null hypothesis if there is a high probability that the difference between the two sample means could have resulted from sampling error; we decide to reject the null hypothesis if there is a low probability that the difference between the two sample means could have resulted from sampling error. A decision to reject the null hypothesis is tantamount to deciding that the two sample means did come from sampling distributions that have different means, and that a real difference exists between the test scores of children taught by Method A and those taught by Method B. Whether the difference in scores is the result of the difference in effectiveness of the two teaching methods or of other factors, such as teacher motivation or environmental differences in the classrooms, will be of concern as the researcher interprets the statistical findings. The statistical test only determines the probability that the findings occurred by chance.

DECISIONS REGARDING THE NULL HYPOTHESIS

Note that the decision is whether or not to reject the null hypothesis. Deciding to not reject the null hypothesis is not the same as deciding to accept it. Why do we make this seemingly minute semantic distinction between "not rejecting" and "accepting" the null hypothesis? We need to remember that the null hypothesis states that there is no difference between the groups. This is an exact statement about the difference between the groups. If the results of our project are inconsistent with this null hypothesis, we can legitimately reject it. However, if our results are consistent with the null hypothesis, they cannot be interpreted as grounds for accepting it, because a finding consistent with the null hypothesis can also be consistent with a number of other hypotheses. For example, if we find that students receiving Method B far outscore the students receiving Method A, we can correctly decide to reject the null hypothesis. On the other hand, if we find that they do not score significantly higher, we cannot legitimately conclude that there is no difference (which would mean accepting the null hypothesis). It may be, in fact, that Method B is a little more effective than Method A (or that Method A is a little better), but that our research project failed to detect this difference. Our decision not to reject the null hypothesis means only that the data obtained in our samples were not sufficiently different to let us conclude that the difference did not occur by chance.

In summary, we should never "accept" the null hypothesis; we should only "reject" or "not reject" it. If a statistical test does not indicate that the two samples came from different populations, we can only conclude that we failed to detect a significant difference. It does not mean that, in fact, no difference exists. Therefore we can only "not reject" the hypothesis that there is no difference. We are not justified in concluding that there *is* no difference.

TESTING THE NULL HYPOTHESIS

In a typical two-sample case, when we test the null hypothesis, we are concerned with the difference between a pair of sample means. By stating the null hypothesis, we make an explicit assumption that the difference between the two sample means is due to sampling error. The next step is to determine the sampling distribution of differences between pairs of sample means.

To show how we decide whether to reject the null hypothesis, we must once again create a hypothetical situation. Suppose we have unlimited time and resources at our disposal and can take many, many samples from the same population. Let's assume that the number of scores in each sample is 50. Earlier we learned that a distribution of these sample means would provide us with the standard error of the mean ($\sigma_{\bar{x}}$). Now we are interested in learning about the sampling distribution of differences between pairs of sample means. To do this, we can imagine that we are able to form every conceivable combination of two sample means in a given population. This will give us an array of pairs of sample means. We then determine the difference between the mean scores for each pair, always subtracting the mean of the second sample from the mean of the first. If we make a distribution of these differences between paired sample means, we will get a sampling distribution that is in the form of a normal distribution. Figure 13-1 shows this sampling distribution of differences between sample means selected from the same population.

An important feature of this distribution of differences is that the mean difference score is always equal to zero. It is evident that the differences between pairs of sample means selected from the same population would vary around zero.

THE STANDARD ERROR OF THE DIFFERENCE BETWEEN MEANS

After computing the differences between the means of a multitude of pairs of samples, we can compute the standard deviation of these differences. The term used to describe the standard deviation of differences between means is the *standard error of the difference*, and its symbol is $\sigma_{\bar{X}_1 - \bar{X}_2}$. In Figure 13-1, the $\sigma_{\bar{X}_1 - \bar{X}_2}$ values are shown along the horizontal axis, exactly as $\sigma_{\bar{X}}$ was in earlier examples. The major distinction between these two measures is that the $\sigma_{\bar{X}}$ is a measure of the dispersion of sample means around the population mean, whereas the $\sigma_{\bar{X}_1 - \bar{X}_2}$ is a measure of the dispersion of differences between paired sample means around a mean of zero.

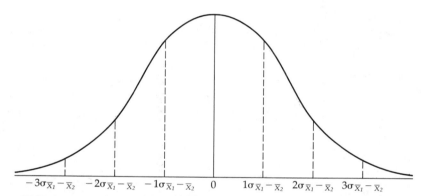

FIGURE 13-1. Sampling distribution of differences between pairs of sample means selected from the same population.

The probability functions of the normal curve apply to this particular sampling distribution of mean differences, because the $\sigma_{\bar{X}_1 - \bar{X}_2}$ is assumed to have been computed from the differences between the means of all possible pairs—that is, the population—of samples. Thus the standard error of the difference is considered a population parameter, and it is assigned the Greek symbol $\sigma_{\bar{X}_1 - \bar{X}_2}$.

How do we use this sampling distribution to decide whether or not to reject the null hypothesis? Using sample data, we can never know for certain whether the null hypothesis is false. Our task is to make an intelligent decision in the face of uncertainty. We can only state the risk we are willing to run of rejecting the null hypothesis when it is really true. Naturally we would like the probability of our making the wrong decision to be quite low.

The individual researcher must determine how much risk of error he is willing to run if he decides to reject the null hypothesis. Of course, he or she must set this risk level before conducting the experiment. Researchers usually set this risk level at $P = .05$ or $P = .01$.

In our example, suppose we have set $P = .05$ as the probability level. In so doing, we are saying that if we decide to reject the null hypothesis, the probability is $P = .05$ that we are making the wrong decision. Since $P = .05$ represents a very small probability of incorrectly rejecting the null hypothesis, we have decided that we are willing to run that much risk. By examining the data in our two samples, we can determine the probability that they came from the same population. If this probability is less than $P = .05$, we reject the null hypothesis that they came from the same population. On the other hand, if the probability of obtaining a difference between the sample means is greater than .05, we do not reject the null hypothesis, and we conclude that the difference between the means could be due to sampling error.

From the properties of the normal curve, we know that .95 of the differences between sample means lie between $-1.96\sigma_{\bar{X}_1-\bar{X}_2}$ and $1.96\sigma_{\bar{X}_1-\bar{X}_2}$. Therefore the probability that the difference between a given pair of sample means lies between these two points is $P = .95$, and the probability that the difference between a pair of means lies outside these two points is $P = .05$ ($P = .025$ below $-1.96\sigma_{\bar{X}_1-\bar{X}_2}$ and $P = .025$ above $1.96\sigma_{\bar{X}_1-\bar{X}_2}$).

After we have distributed all of the differences between the paired sample means in our hypothetical situation, suppose we determine that the $\sigma_{\bar{X}_1-\bar{X}_2}$ of this distribution is 3 points. To determine the two values that correspond to $-1.96\sigma_{\bar{X}_1-\bar{X}_2}$ and $1.96\sigma_{\bar{X}_1-\bar{X}_2}$, we calculate as follows:

$$\text{For } -1.96\sigma_{\bar{X}_1-\bar{X}_2}: \quad -1.96 \times 3 = -5.88$$
$$\text{For } 1.96\sigma_{\bar{X}_1-\bar{X}_2}: \quad 1.96 \times 3 = 5.88$$

Figure 13-2 presents the sampling distribution of differences between sample means, with these two values shown as cutoff points for $P = .95$. Note that, for these data, the mean of the distribution is zero, with the differences between the means distributed on either side of this midpoint. (The negative differences are results of our method of pairing sample means, because in some pairs we are subtracting a larger mean from a smaller one. This occurs in 50% of the pairs, so that half of the distribution of differences consists of negative differences and half consists of positive differences.)

Since we originally decided to set $P = .05$ as our probability level, Figure 13-2 indicates the area under the curve where an obtained difference between a pair of sample means must lie if we are to reject the null hypothesis. We can state that the probability, due to sampling error only, of obtaining two samples whose means differ by 5.88 points or less (regardless of whether the difference is negative or positive) is $P = .95$. The probability of obtaining two samples

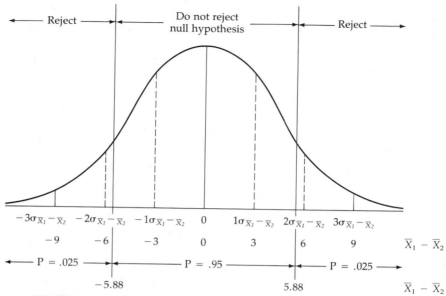

FIGURE 13-2. Sampling distribution of differences between pairs of sample means selected from the same population.

whose means differ by more than 5.88 points is $P = .05$. Looking back at the data in our two samples, we find that the mean score of the students given Method B was 4 points higher than the mean score of the students given Method A. The sampling distribution in Figure 13-2 reveals that a mean difference of 4 points falls within the $P = .95$ area, indicating that such a difference has a high probability of occurring by chance. Since the probability is not less than .05 that the obtained difference between means occurred by chance, we cannot reject the null hypothesis that there is no difference in the effectiveness between Method A and Method B for teaching music appreciation to fourth-grade students. We must conclude that the difference between the means of our two samples could have been due to sampling error. On the other hand, if the difference between our sample means had been 8 points, for instance, we would have rejected the null hypothesis, because the probability of our obtaining such a large difference solely due to sampling error is less than .05.

We have shown in Figure 13-2 the sampling distribution of differences between a multitude of pairs of sample means to illustrate how we can make probability statements about their occurrence and how we can reach decisions regarding the null hypothesis based on the probability that a pair of sample means will differ because of sampling error. In actuality, we never have the luxury of many pairs of samples; research studies usually deal with only two samples. The statistical problem in such situations is to use the values of the samples themselves to estimate the amount of variability in the distribution of differences between sample means. From this estimate we can form a sampling distribution and determine the probability that the two sample means are from the same population.

Fortunately there is a method for estimating the standard error of the difference. An estimate of this value is obtained by applying a statistical formula to the data in the two samples. The process by which we estimate the standard error of the mean and develop sampling distributions from this estimate is discussed in the next chapter.

EXERCISES

1. A researcher wants to determine whether the variable method and the constant method of training chickens to peck at a red circle differ in effectiveness. What is the research hypothesis he wants to test?
2. State the null hypothesis that is to be tested for the research hypothesis in Exercise 1.
3. Suppose a multitude of pairs of samples have been randomly selected, where each sample in a pair has been given a different method of training. Suppose that $\sigma_{\bar{X}_1 - \bar{X}_2} = 4$. If the researcher set $P = .05$ as the probability level and if he obtained a mean of 40 for the sample given the variable method and a mean of 50 for the sample given the constant method, would he reject the null hypothesis?
4. Given the data in Exercise 3, would the researcher reject the null hypothesis if he obtained a mean of 48 for the variable-method sample and a mean of 42 for the constant-method sample?
5. A statistics instructor has a hunch that the exercises in Group A and Group B are not of equal difficulty. What is the research hypothesis he wants to examine?
6. State the null hypothesis that is to be tested for the research hypothesis given in Exercise 5.
7. Suppose a multitude of pairs of samples have been randomly selected, where each sample in a pair has been given a different set of exercises. Suppose $\sigma_{\bar{X}_1 - \bar{X}_2} = 2.5$ points. If the instructor sets $P = .05$ as the probability level and if he obtains a sample mean for Group A exercises of 12 points and a sample mean for Group B exercises of 8 points, would he reject the null hypothesis?
8. Given the data in Exercise 7, would the instructor reject the null hypothesis if he obtains a sample mean for Group A exercises of 7 points and a sample mean for Group B exercises of 14 points?

14 TESTING FOR THE DIFFERENCE BETWEEN POPULATION MEANS

A typical research situation is one in which we want to decide whether or not to reject the null hypothesis when we have only the data in two samples. Since we do not have an infinite number of pairs of samples that will enable us to form a sampling distribution of differences, we must use statistical methods to estimate the value of the standard error of the difference. Using this estimate, we can then specify a sampling distribution to use in examining the null hypothesis.

This chapter presents three methods for estimating the standard error of the difference based on the data in two samples. Each method assumes that the scores in the samples come from a normally distributed population. The symbol for an estimate of the population standard error of the difference, calculated from sample data, is $s_{\bar{X}_1 - \bar{X}_2}$. The statistical test used to analyze the difference between a pair of sample means, when $s_{\bar{X}_1 - \bar{X}_2}$ is used to specify the sampling distribution, is commonly called a t test, because distributions of this sort belong to the family of t distributions. The first method involves calculating $s_{\bar{X}_1 - \bar{X}_2}$ when the data are obtained from two independent samples and the variances in the respective populations are assumed to be equal. The second method is also for independent samples, but where the variances are assumed to be unequal. The third method uses data from two dependent samples.

Two samples are considered independent if members have been randomly assigned to the two groups. When this procedure is followed, the scores obtained by members of one sample are not in any way related to or influenced by the scores obtained by members of the other sample. Thus the scores in the two samples are independent of one another, and the means of the two samples are said to be *independent sample means*. The example in Chapter 13 in which fourth-grade children were randomly assigned to the two methods of teaching music appreciation involved independent samples because the assignment of one child to a particular sample did not influence the assignment of other children to the samples. Therefore the means of these two sets of scores represent independent sample means.

However, the means are dependent, or related, if the scores in the two samples are obtained from the same sample of individuals. This occurs, for example, when the individuals in one sample receive both a pretest and a posttest. Since in this case the same individuals produced both sets of scores, we cannot say that the posttest scores are unrelated to, or are independent of, the pretest scores. Therefore we say that the means of these two sets of data represent *dependent sample means*.

Samples also cannot be considered independent if the individuals in the two samples have been paired in any way. For example, suspecting that the intelligence levels of pupils affect their performance on a test of manual dexterity, a researcher may decide to group pupils into pairs on the basis of their IQs, and then randomly divide each pair between the two samples. The mean manual dexterity scores obtained from these two samples are dependent because the individuals in the samples have been assigned to groups on the basis of their IQs, with the assignment of one member of a pair to one group dictating the assignment of the other member to the second group. Thus, in this case, the study involves related, or dependent, sample means, even though the two samples are made up of different individuals.

We use different statistical formulas to calculate $s_{\bar{X}_1 - \bar{X}_2}$ depending on whether we are dealing with independent or dependent sample means. First we examine methods for testing the null hypothesis when the data represent independent sample means; later we present methods for analyzing dependent means.

TESTING THE DIFFERENCE BETWEEN INDEPENDENT MEANS—VARIANCES ASSUMED EQUAL

Mathematical statisticians have given us a set of formulas that permits us to calculate an estimate of the standard error of the difference when we are given only the data in two independent samples. Two methods for estimating the standard error of the difference are presented here. The appropriate one to use depends on whether we can assume that the variances in the respective populations are equal or not. We really have no adequate statistical test to apply to help us decide this. Fortunately we are on rather safe ground assuming the equality of the population variances when we have equal-sized samples. That is, relatively large differences in variance estimates where equal-sized samples are involved do not seriously affect the conclusions made on the basis of a *t* test.

However, when variance estimates are quite unequal and the sample sizes differ markedly, the effects on the conclusions can be quite dramatic. In such cases, we need to use a *t* test that does not require us to assume equal variances in the populations. We present both methods for performing a *t* test in the next two sections.

Suppose we want to determine whether or not to reject the null hypothesis that Method 1 and Method 2 are equally effective for teaching reading. If we assume that the null hypothesis is true, we are saying that the reading scores in the two samples all come from the same population of reading scores. If we also assume that the population variances are equal, our best estimate of the population variance should be obtained by pooling the sums of squares and the degrees of freedom that we obtain from both samples. Formula 26 gives us two equivalent methods for estimating the population variance s^2 by pooling the data in the two samples.

Formula 26a uses the sums of squares in both samples; Formula 26b uses variance estimates derived from each of the samples. Both formulas yield identical estimates of the population variance. With the estimate of s^2 obtained from Formula 26, we can use Formula 27 to estimate the standard error of the difference.

FORMULA 26

Estimate of the common population variance (pooled variance) from the data in two samples. (Formulas 26a and 26b are equivalent.)

$$s^2 = \frac{\Sigma x_1^2 + \Sigma x_2^2}{N_1 + N_2 - 2} \quad \text{(Formula 26a)}$$

$$s^2 = \frac{(N_1 - 1)s_1^2 + (N_2 - 1)s_2^2}{N_1 + N_2 - 2} \quad \text{(Formula 26b)}$$

FORMULA 27

Estimate of the standard error of the difference between means (pooled-variance method).

$$s_{\bar{X}_1 - \bar{X}_2} = \sqrt{\frac{s^2}{N_1} + \frac{s^2}{N_2}}$$

FORMULA 28

Calculation of the t ratio for independent means.

$$t = \frac{\bar{X}_1 - \bar{X}_2}{s_{\bar{X}_1 - \bar{X}_2}}$$

$$df = N_1 + N_2 - 2$$

We can now use $s_{\bar{X}_1 - \bar{X}_2}$ to develop a sampling distribution of differences between pairs of sample means for use in determining the probability that the obtained difference between the two sample means occurred by chance. Since this sampling distribution is derived from an estimate of the standard error of the difference ($s_{\bar{X}_1 - \bar{X}_2}$) rather than from a known population parameter ($\sigma_{\bar{X}_1 - \bar{X}_2}$), it follows the appropriate t distribution rather than the normal distribution as was the case in Chapter 10. The t distribution that is appropriate to represent the sampling distribution depends on the degrees of freedom associated with the two sample means. For Sample 1 the df is $N_1 - 1$, and for Sample 2 the df is $N_2 - 1$. Thus, assuming the variances are equal, the pooled df of the samples is $(N_1 - 1) + (N_2 - 1)$ or $N_1 + N_2 - 2$.

Once we have calculated $s_{\bar{X}_1 - \bar{X}_2}$ and have identified the appropriate t distribution based on the pooled df, we are in a position to develop the sampling distribution and to evaluate the difference between the two obtained sample means.

Formula 28 gives the method for calculating the t ratio for independent means, under the assumption of equal variances. Formula 28 shows that we compute a t ratio based on the ratio of the difference between sample means to the estimate of the standard error of the difference. If this obtained t ratio is larger than we expect, based on our preset level for significance, we reject the null hypothesis. If the t ratio is smaller, we do not reject the null hypothesis.

Figure 14-1 depicts the sampling distribution curve in the form of a t distribution. The curve of the distribution is indicated by a dotted line because the shape of the particular t distribution depends on the degrees of freedom associated with it. Likewise, the vertical lines indicating the .95 area under the curve are dotted because their placement depends on the particular t distribution. For t distributions with small dfs, these lines are located farther out on the horizon-

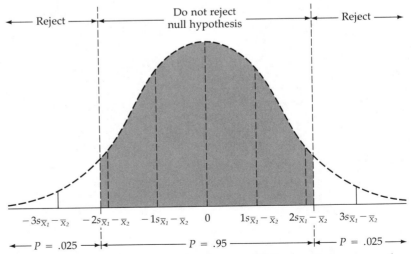

FIGURE 14-1. The sampling distribution of differences between sample means (t distribution).

tal axis—that is, farther apart—than they are for t distributions with large dfs. For example, Table 2 indicates that for $df = 17$, at $P = .05$, these vertical lines would be located at $t = -2.11$ and $t = 2.11$. For $df = 5$, at $P = .05$, they would be located at $t = -2.571$ and $t = 2.571$. For an infinite df, the t distribution is identical to the normal distribution, and the vertical lines for $P = .05$ are located at $t = -1.96$ and $t = 1.96$; these values are the same as the z scores delineating the same area, as we learned when examining the properties of the normal curve.

EXAMPLE 14-1 An educator wanted to find out if different methods of driver training resulted in different driver proficiency levels. He randomly assigned students to two groups and taught one group using videotape demonstrations (Method 1) and the other using live demonstrations (Method 2). He set $P = .01$ as his level of significance. At the conclusion of the training, he obtained the following driver proficiency scores for the students in the two groups.

Method 1		Method 2	
27	19	24	31
28	25	28	29
20	31	31	26
27	24	27	33
30	23	28	29
28	27	23	28
		29	26

• • • • • • •

The information given in Example 14-1 indicates that the educator is interested only in detecting a difference, regardless of whether the difference favors Method 1 or Method 2. By setting $P = .01$ as his significance level, he has decided that he will not reject the null hypothesis if the difference between his two sample means lies in the probability area $P = .99$ of the appropriate sampling distribution. On the other hand, he will reject the null hypothesis if the difference between the means has a probability of occurring by chance as low as $P = .01$. Since he is willing to reject the null hypothesis regardless of the direction of the difference, he is interested in differences that fall in either tail of the sampling distribution. This means that his significance level of $P = .01$ must be divided between the two tails, with $P = .005$ in either tail. This is called a two-tail test. The distinction between one-tail and two-tail tests is examined in depth in Chapter 15.

To test the null hypothesis that there is no difference between the driver proficiency scores of students taught by Method 1 and by Method 2, he must determine the means of the two samples, calculate the sum of squares for each sample, estimate the population variance from the data in the two samples, and estimate the standard error of the difference. These calculations are as follows:

Method 1	Method 2
$N_1 = 12$	$N_2 = 14$
$\Sigma X_1 = 309$	$\Sigma X_2 = 392$
$\Sigma X_1^2 = 8,107$	$\Sigma X_2^2 = 11,072$
$\bar{X}_1 = \dfrac{309}{12} = 25.75$	$\bar{X}_2 = \dfrac{392}{14} = 28.0$
$s_2^2 = 13.659$	$s_1^2 = 7.385$

Assume that the variances are equal. Therefore, we may use the "pooled variance" method for estimating the standard error of the difference.

Using Formula 7b:
$$\Sigma x_1^2 = \Sigma X_1^2 - \frac{(\Sigma X_1)^2}{N_1} = 8,107 - \frac{(309)^2}{12} = 150.25$$

$$\Sigma x_2^2 = \Sigma X_2^2 - \frac{(\Sigma X_2)^2}{N_2} = 11,072 - \frac{(392)^2}{14} = 96.0$$

Using Formula 26a:
$$s^2 = \frac{\Sigma x_1^2 + \Sigma x_2^2}{N_1 + N_2 - 2} = \frac{150.25 + 96.0}{12 + 14 - 2} = 10.26$$

Using Formula 27:
$$s_{\bar{X}_1 - \bar{X}_2} = \sqrt{\frac{s^2}{N_1} + \frac{s^2}{N_2}} = \sqrt{\frac{10.26}{12} + \frac{10.26}{14}} = 1.26$$

The degrees of freedom associated with the two independent sample means are $N_1 + N_2 - 2 = 24$. Using this value and Table 2, the educator can form the appropriate sampling distribution. Table 2 shows that for $df = 24$, the t values that cut off $P = .01$ of the area ($P = .005$ in each tail) are $t = -2.797$ and $t = 2.797$. The appropriate sampling distribution for these data is shown in Figure 14-2.

He must now determine where the difference between the two sample means lies, in relation to the regions representing rejection of the null hypothesis. For this, he needs to compute the t ratio, using Formula 28.

$$t = \frac{\bar{X}_1 - \bar{X}_2}{s_{\bar{X}_1 - \bar{X}_2}} = \frac{25.75 - 28.0}{1.26} = -1.79$$

Figure 14-2 shows us that $t = -1.79$ falls in the area between $t = -2.797$ and $t = 2.797$, which is our do-not-reject region. Therefore, he must conclude that the difference between the two sample means is not large enough for him to reject the null hypothesis that they came from the same population. In this study, the educator must conclude that the obtained difference in mean scores between the students taught by Method 1 and those taught by Method 2 could have occurred by chance.

Of course, to perform this statistical test, it is not necessary to display the sampling distribution graphically, as we did in Figure 14-2. We drew the curve merely to illustrate the process by which we make decisions using the underly-

TESTING FOR THE DIFFERENCE BETWEEN POPULATION MEANS

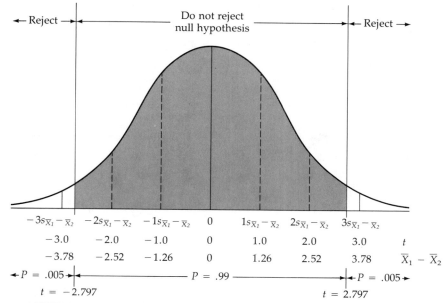

FIGURE 14-2. The sampling distribution of differences for data in Example 14-1 (t distribution for $df = 24$).

ing sampling distribution. In practice, we determine the degrees of freedom associated with the data in our study and use Table 2 to determine the value of t that is needed at a given significance level. We then compute the t ratio, using the appropriate formulas, and compare it with the tabled value. If the obtained t ratio is larger than the tabled t value, we reject the null hypothesis. If it is less than the tabled t value, we do not reject the null hypothesis.

TESTING THE DIFFERENCE BETWEEN INDEPENDENT SAMPLE MEANS—VARIANCES NOT ASSUMED EQUAL

The same logic prevails for performing the t test of independent sample means in cases where the population variances cannot be assumed equal, except that we are not permitted to pool the variance estimates in deriving the standard error of the difference. Instead, we must use Formula 29 in estimating the standard error of the difference. The t ratio is computed in the same manner as before, as shown in Formula 30, but the degrees of freedom to be used in evaluating the significance of the t ratio are determined by a rather complex formula, developed by Welch[1], and presented in Formula 30. Example 14-2 illustrates the use of these formulas.

[1] Welch, B. L., "The Significance of the Difference Between Two Means When the Population Variances Are Unequal," *Biometrika* 29 (1938): pp. 350–62.

FORMULA 29 Estimate of the standard error of the difference between means (variances assumed unequal).

$$s_{\bar{X}_1 - \bar{X}_2} = \sqrt{\frac{s_1^2}{N_1} + \frac{s_2^2}{N_2}}$$

FORMULA 30 Calculation of the t ratio for independent means (variances assumed unequal).

$$t = \frac{\bar{X}_1 - \bar{X}_2}{s_{\bar{X}_1 - \bar{X}_2}}$$

$$df = \frac{(s_{\bar{X}_1}^2 + s_{\bar{X}_2}^2)^2}{(s_{\bar{X}_1}^2)^2/(N_1 + 1) + (s_{\bar{X}_2}^2)^2/(N_2 + 1)} - 2$$

EXAMPLE 14-2 An insurance agency personnel director wanted to determine if male and female employees differed in the amount of time they spent speaking to clients on the telephone. Unfortunately there were few males employed in the agency. However, she selected a random sample of each sex. She set $P = .05$ as her level of significance and obtained the following times in minutes:

Males		Females	
15	11	12	15
13	16	25	28
16	17	16	10
19		18	17
		10	22
		34	32
		20	27
		24	

The sample statistics for these two samples are as follows:

	Males	Females
N	7	15
Mean	15.286	20.667
Standard deviation	2.628	7.603
Variance	6.905	57.810
Standard error of mean	.993	1.963

• • • • • • •

Note that the sample sizes differ markedly. Also, the estimated variances for these samples show a much larger variance estimate for the female sample than for the male sample. This indicates that a test should be used that does not assume that the variances in the populations are equal. Therefore we should use the Formulas 29 and 30 to calculate the *t* ratio for independent means.

The standard error of the difference for these data is computed using Formula 29, as follows:

$$s_{\bar{X}_1 - \bar{X}_2} = \sqrt{\frac{6.905}{7} + \frac{57.810}{15}} = 2.200$$

Substituting the values in Formula 30 yields the *t* ratio.

$$t = \frac{15.286 - 20.667}{2.2} = -2.446$$

The degrees of freedom are determined by using the procedure in Formula 30.

$$df = \frac{[(.993)^2 + (1.963)^2]^2}{[(.993)^2]^2/(7 + 1) + [(1.963)^2]^2/(15 + 1)} - 2 = 20$$

To determine whether or not to reject the null hypothesis that there is no difference in the amount of time spent on the telephone, the personnel director must evaluate the *t* ratio of -2.446 with 20 degrees of freedom. This is a two-tailed test. Using Table 2, at $P = .05$ we see that a *t* ratio of 2.086 or larger is needed to reject the null hypothesis. Since her obtained *t* ratio exceeds this amount, she rejects the null hypothesis and concludes that female employees speak longer on the phone than do male employees.

TESTING THE DIFFERENCE BETWEEN DEPENDENT MEANS

The means of samples are dependent, or related, in studies in which the same individuals are tested at two different times or in which the individuals selected for the two samples have been paired in some way, such as matching them in terms of a variable that could affect their scores. In these studies, we can compute the difference between each pair of scores and use these differences to estimate the population standard error of the mean difference scores, which is represented by $s_{\bar{D}}$.

To compute this estimate, we must first obtain an estimate of the population variance of the difference scores, using Formula 31.

Formula 31 requires that we compute the sum of the difference scores (ΣD), where $D = X_1 - X_2$ for each of the N pair of scores and also compute the sum of the squares of the difference scores. Note that ΣD^2 is not the same as $(\Sigma D)^2$. For ΣD^2, we square each difference score before we sum them; for $(\Sigma D)^2$, we sum the difference scores and then square this sum. In Formula 31, N indicates the number of difference scores.

FORMULA 31 — Estimate of the population variance of difference scores.

$$s_D^2 = \frac{N\Sigma D^2 - (\Sigma D)^2}{N(N-1)}$$

where: $D = X_1 - X_2$ for each pair of scores

Formula 32 is then used to estimate the population standard error of the mean difference scores.

FORMULA 32 — Estimate of the population standard error of the mean difference scores. (Formulas 32a and 32b are equivalent.)

$$s_{\bar{D}} = \sqrt{\frac{s_D^2}{N}} \quad \text{(Formula 32a)}$$

$$s_{\bar{D}} = \sqrt{\frac{N\Sigma D^2 - (\Sigma D)^2}{N^2(N-1)}} \quad \text{(Formula 32b)}$$

where: N = number of pairs of scores

There are two equivalent formulas used to calculate this estimate. Formula 32a uses s_D^2 in its calculation; Formula 32b uses the difference scores. The t ratio is then computed using Formula 33, in which the degrees of freedom associated with the t distribution is the number of pairs of scores minus one.

FORMULA 33 — Calculation of the t ratio for nonindependent means.

$$t = \frac{\bar{X}_1 - \bar{X}_2}{s_{\bar{D}}}$$

$df = N - 1$ pairs of scores

We will use Example 14-3 to illustrate the computational procedure for testing the difference between dependent means.

EXAMPLE 14-3 A high school typing instructor wanted to determine if students taking a typing proficiency test would score differently depending on whether the test was administered in the morning or the afternoon. She randomly selected ten students from her typing class and gave each student two typing tests, one in the morning and one in the afternoon. To overcome any bias due to the order of testing, she alternated the sequence of the testing by giving half of the sample the morning test first and the other half the afternoon test first. She set $P = .05$ as the level of significance and obtained the following typing proficiency scores.

Student	Morning test scores (X_1)	Afternoon test scores (X_2)	D	D^2
A	18	16	−2	4
B	19	19	0	0
C	17	16	−1	1
D	22	18	−4	16
E	15	17	+2	4
F	16	15	−1	1
G	18	14	−4	16
H	19	12	−7	49
I	13	10	−3	9
J	20	17	−3	9
$N = 10$	$\Sigma X_1 = 177$	$\Sigma X_2 = 154$	$\Sigma D = -23$	$\Sigma D^2 = 109$

· · · · · · ·

In Example 14-3, ΣD and ΣD^2 have already been calculated. The estimate of the population variance of morning/afternoon difference scores and the estimate of the population standard error of the mean difference scores are as follows:

$$\bar{X}_1 = \frac{177}{10} = 17.7 \qquad \bar{X}_2 = \frac{154}{10} = 15.4$$

Using Formula 31: $\quad s_D^2 = \dfrac{N\Sigma D^2 - (\Sigma D)^2}{N(N-1)} = \dfrac{10(109) - (-23)^2}{10(10-1)} = 6.23$

Using Formula 32a: $\quad s_{\bar{D}} = \sqrt{\dfrac{s_D^2}{N}} = \sqrt{\dfrac{6.23}{10}} = .79$

Using Formula 32b: $\quad s_{\bar{D}} = \sqrt{\dfrac{N\Sigma D^2 - (\Sigma D)^2}{N^2(N-1)}}$

$$= \sqrt{\dfrac{10(109) - (-23)^2}{(10)^2(9)}} = .79$$

The degrees of freedom associated with nonindependent sample means are $N - 1$. Thus, in Example 14-3, the degrees of freedom are $10 - 1 = 9$. Table 2 shows that the values of t that designate $P = .05$ in the tails of the curve ($P = .025$ in each tail) are $t = -2.262$ and $t = 2.262$. The appropriate sampling distribution for this example is the t distribution based on 9 degrees of freedom; it is shown in Figure 14-3.

The final step in this procedure is to use Formula 33 to determine the t ratio based on the difference between the two sample means.

Using Formula 19: $\quad t = \dfrac{\bar{X}_1 - \bar{X}_2}{s_{\bar{D}}} = \dfrac{17.7 - 15.4}{.79} = 2.91$

Because the obtained t ratio is larger than the tabled value needed for significance at $P = .05$, it falls in the rejection region, as Figure 14-3 shows.

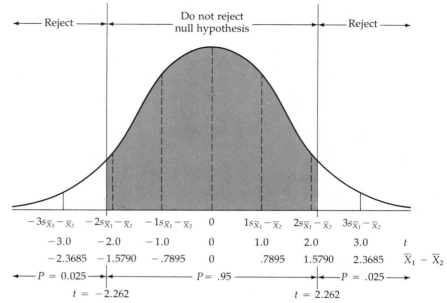

FIGURE 14-3. The sampling distribution of differences for data in Example 14-3 (t-distribution for $df = 9$).

Therefore, the typing instructor should reject the null hypothesis that there is no difference in students' typing proficiency scores when they are tested in the morning and in the afternoon. The data indicate that students score significantly higher when they are tested in the morning than they do when they are tested in the afternoon. Of course, since the instructor has rejected the null hypothesis at the .05 significance level, she must realize that there is this small a probability that the mean difference she obtained between the two sets of data occurred by chance and was not the result of the different times of testing. However, since she can never prove that there is a real difference, she decides to settle for this quite low probability of her being wrong.

In this chapter, we have discussed methods for determining whether the observed difference between the means of two samples is a significant one or whether there is a high probability that the difference is merely the result of sampling error. The procedures leading to the decision to reject or not reject the null hypothesis are essentially these:

1. A research hypothesis is stated regarding the difference between the means of two populations of scores.
2. The research hypothesis is reworded as a null hypothesis, which states that there is no difference between the population means.
3. The probability level of significance is decided. This level represents the risk we are willing to take of being incorrect if we decide to reject the null hypothesis.
4. Two sets of sample data are collected and their means are computed.

5. Assuming that the null hypothesis is true, the appropriate t distribution is specified to serve as the sampling distribution of the differences between pairs of sample means.
6. A t ratio based on the difference between the obtained sample means and the estimate of the standard error of the difference between the means is computed.
7. The probability of occurrence of the t ratio within the sampling distribution is determined.
8. If the probability of the occurrence of the t ratio is less than the pre-specified significance level, the null hypothesis is rejected; if it is greater, the null hypothesis is not rejected.

EXERCISES

For these exercises, when using Table 2, assume all exercises to be for a nondirectional hypothesis (two-tail test).

1. A swimming instructor wanted to determine if there was a difference in the effectiveness of two methods of instruction in distance swimming. He randomly selected two groups of nonswimmers and gave one group the intensive instructional method and the other group the periodic instructional method. He then tested each group on swimming ability and obtained the following data. He set $P = .05$ as the level of significant. Should he reject the null hypothesis?

Intensive method			Periodic method		
45	41	44	43	52	48
37	37	38	44	55	51
40	39	41	48	50	45
39	42	36	54	53	49
37	42	35	45	49	47
40	41	39	46	47	49
38	42	37	47	48	48

2. A test developer wanted to determine if the two forms of her intelligence test were comparable. She randomly selected a group of students and administered both Form A and Form B to each student. She obtained the following test data and set $P = .01$ as the level of significance. Should she reject the null hypothesis?

Student	Form A	Form B	Student	Form A	Form B
A	104	102	K	99	102
B	96	96	L	100	104
C	106	105	M	97	96
D	101	102	N	98	97
E	98	94	O	94	99
F	99	98	P	101	102
G	111	108	Q	99	98
H	102	103	R	103	104
I	100	94	S	96	98
J	97	96	T	100	100

3. An experimental psychologist was interested in determining the relative effects of two drugs on guinea pigs. She randomly selected two samples and injected one sample with drug XX-7 and the other sample with drug XX-8. Drug XX-8 is a very expensive drug and she was unable to test many guinea pigs with it. After the injections, she gave the members of each group a pain-sensitivity test and obtained the following results. She set $P = .05$ as the level of significance. Should the null hypothesis be rejected?

Drug XX-7			Drug XX-8		
79	77	78	98	92	85
78	79	81	98	90	92
80	83	82	99	94	91
83	79	81			
79	80	80			
78	82	78			

15 MAKING DECISIONS ABOUT THE NULL HYPOTHESIS

A researcher embarking on a particular research project does not work in a vacuum. Much thought and effort must be expended before we can formulate the research hypothesis, select the appropriate research design, and specify the procedures and techniques for data gathering, analysis, and interpretation.

Our first task is the formulation of the research hypothesis. This hypothesis is usually the outgrowth of considerable study and it reflects our desire to answer a question about the relationship between variables. We conduct our experiment for the express purpose of obtaining objective evidence to corroborate or refute certain aspects of existing theory or to verify relationships between variables that will lead to the development of new theoretical positions.

Of course, theory development depends on the continued confirmation of hypotheses, rather than on the results of a single isolated experiment. No scientist would be foolhardy enough to develop a theoretical position based on the statistical analysis of one research project!

As we have seen earlier, we are not in a position to test a research hypothesis directly. We must first convert it to a null hypothesis and then, using the sample statistics, we must decide whether or not to reject it. Because we must depend on sample data to substantiate or refute the hypothesis, we know that we cannot make either decision with certainty. In this process, there is always a chance that the sample data will lead us to an incorrect decision. In fact, the results of a single experiment must be viewed with a degree of skepticism, and they should not form the basis of any definitive judgments about the truth of the hypothesis.

This point is made to emphasize the fact that, although we say that we "reject" or "do not reject" the null hypothesis on the basis of sample data, this does not mean that we have irrevocably decided that the hypothesis is or is not true. The phrase "reject the null hypothesis" should be taken to mean: "The statistical analysis of the data obtained from the samples in this experiment indicates that there is such a low probability that they came from the same population that not to reject such a notion would be flying in the face of obtained evidence." Another way of stating the phrase "not reject the null hypothesis" is: "The statistical analysis of the data obtained from this experiment indicates that there is a high probability that the observed difference between the samples could have occurred through sampling error, and that without further evidence, it would be foolish to conclude that the difference was due to some factor other than chance." Although the terms "reject" and "not reject" are used for convenience throughout this book, they should always be interpreted in this manner.

In this chapter we examine the manner in which a research hypothesis is stated and the effect it has on our decision regarding the null hypothesis. We also discuss how significance levels are set and the types of errors that are likely to occur when we make decisions about the null hypothesis.

ONE- AND TWO-TAIL TESTS

Sometimes we make a hypothesis stating that one experimental treatment is more effective than another treatment. At other times we do not state which treatment we think is more effective than the other but merely hypothesize that there is a difference in effectiveness between the two treatments. The type of research hypothesis that is made determines how the significance of the difference between the treatments is evaluated.

In either case, the statistical hypothesis used in testing is the null hypothesis. As we stated earlier, a null hypothesis indicating that there is no difference between the means of two populations is symbolized by:

$$H_0: \quad \mu_1 - \mu_2 = 0$$

In the statistical analysis of data, the research hypothesis is considered an alternate to the null hypothesis. We designate the alternate hypothesis H_1. If the alternate hypothesis is: "There is a difference in effectiveness between Method 1 and Method 2 in teaching reading to third-grade pupils" it can be expressed as:

$$H_1: \quad \mu_1 \neq \mu_2$$

This expression tells us that the population mean of third-grade pupils given Method 1 is not equal to the population mean of comparable pupils given Method 2. In this case, the alternate hypothesis does not state the direction of the difference—that is, which method is superior. It only states that there is a difference. Therefore the rejection region for the null hypothesis is located in both tails of the sampling distribution, with one tail indicating that Method 1 is superior to Method 2 and with the other tail indicating the opposite possibility. We reject the null hypothesis if we obtain a difference between the sample means resulting in a test statistic that is located in either rejection region of the sampling distribution.

The examples of hypothesis testing procedures presented in Chapters 13 and 14 were all concerned with testing nondirectional alternative hypotheses, although we did not explicitly say so at that time. In all of those examples, the area under the curve representing the significance level, such as $P = .05$, was equally divided between the two tails of the sampling distribution, with half of it ($P = .025$) located in each tail. When the rejection regions are separated in this manner, we are making a *two-tail test*.

The way in which the research hypothesis is stated dictates the nature of H_1. If the research hypothesis is nondirectional, H_1 is symbolized by $H_1:\ \mu_1 \neq \mu_2$ and a two-tail test is made. If the research hypothesis indicates which population mean is thought to be superior to the other, the alternate hypothesis is a directional one. For example, if the alternate hypothesis states that Method 1 is superior to Method 2, the direction of the difference is indicated, and the hypothesis is expressed as:

$$H_1:\ \mu_1 > \mu_2$$

(The symbol $>$ means "greater than," and the symbol $<$ means "less than.") When the hypothesis is a directional one, the rejection region is located entirely in one tail of the sampling distribution, and we make a *one-tail test*. In such cases we reject the null hypothesis only if we obtain a difference between sample means that lies in the tail of the sampling distribution that represents the rejection region. The problem of determining whether a research hypothesis should be directional or nondirectional involves a number of considerations that we will take up later in this chapter.

EXAMPLE 15-1 An educational products developer has produced a videotape that he claims is superior for teaching basic geometric patterns to first-grade pupils to the conventional flash-card method. Before deciding to purchase this videotape, a school principal decided to test its effectiveness. He hypothesized: "Children will score higher on a test of basic geometric patterns when they are taught by the videotape method (Method 1) than when they are taught by the conventional flash-card method (Method 2)." He set $P = .05$ as his level of significance. He randomly selected two groups of first-grade pupils, assigned each group one of the methods, and obtained the following test data:

Method 1	Method 2
$N_1 = 12$	$N_2 = 14$
$\bar{X}_1 = 40$	$\bar{X}_2 = 37.5$
$\Sigma x_1^2 = 125$	$\Sigma x_2^2 = 110$

• • • • • • •

In Example 15-1, the developer rejects the null hypothesis only if the videotape presentation (Method 1) is superior. His alternate hypothesis is a directional one, so he will make a one-tail statistical test. The hypotheses for this example are:

$$H_0: \mu_1 = \mu_2$$
$$H_1: \mu_1 > \mu_2$$

He rejects H_0 and accepts H_1 only if $\bar{X}_1$ is sufficiently larger than $\bar{X}_2$. If $\bar{X}_2$ is larger than $\bar{X}_1$, then, of course, H_1 cannot be substantiated, and he need not proceed with the test of significance. If $\bar{X}_1$ is larger than $\bar{X}_2$, the difference lies in the hypothesized direction, and he needs to specify the sampling distribution and to ascertain whether the difference between $\bar{X}_1$ and $\bar{X}_2$ is large enough to permit him to reject the null hypothesis at his preset significance level. In this example, we assume that the variances are equal. If he assumes that the data came from normally distributed populations, a one-tail t test for independent means is appropriate for testing the null hypothesis. For the data in Example 15-1, the t ratio is calculated as follows:

Using Formula 26a: $$s^2 = \frac{\Sigma x_1^2 + \Sigma x_2^2}{N_1 + N_2 - 2} = \frac{125 + 110}{12 + 14 - 2} = 9.79$$

Using Formula 27: $$s_{\bar{X}_1 - \bar{X}_2} = \sqrt{\frac{s^2}{N_1} + \frac{s^2}{N_2}} = \sqrt{\frac{9.79}{12} + \frac{9.79}{14}} = 1.23$$

Using Formula 28: $$t = \frac{\bar{X}_1 - \bar{X}_2}{s_{\bar{X}_1 - \bar{X}_2}} = \frac{40 - 37.5}{1.23} = 2.03$$

$$df: \quad N_1 + N_2 - 2 = 12 + 14 - 2 = 24$$

The next task is to determine the appropriate sampling distribution and to specify the rejection region for the null hypothesis. Because the developer has made a directional hypothesis, he must designate $P = .05$ in one tail of the sampling distribution, rather than splitting it between the two tails as we did when we were making a two-tail test. Table 2 may be used to determine the critical t values for directional (one-tail) and for nondirectional (two-tail) tests. Table 2 has two sets of column headings: the top row gives P values for directional tests and the second row gives P values for nondirectional tests. Note that the probability levels for two-tail tests are twice as large as those for one-tail tests. This is logical because in two-tail tests we are placing the rejection region in both tails rather than just in one. When using Table 2, you must first determine which set of column headings is appropriate for the particular type of hypothesis you are testing. To illustrate the use of Table 2, for 10 degrees of freedom, if the significance level is set at $P = .05$, a one-tail test requires a t ratio of 1.812, whereas a two-tail test requires a t ratio of 2.571.

For Example 15-1, where $df = 24$, the t value designating $P = .05$ in one tail of the sampling distribution is $t = 1.711$. The sampling distribution appropriate to this example is shown in Figure 15-1. Note that in the figure the entire rejection region ($P = .05$) is placed in the right tail of the sampling distribution. Therefore only positive t ratios (those favoring Method 1 over Method 2) at

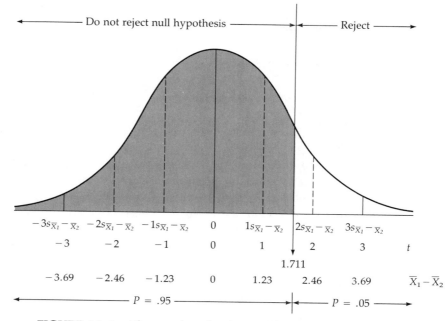

FIGURE 15-1. The sampling distribution of differences between means for the data in Example 15-1 (t distribution for $df = 24$).

or exceeding $t = 1.711$ lead to the rejection of the null hypothesis. The data in Example 15-1 yielded a t ratio of 2.03. This value exceeds the critical value $t = 1.711$ and thus lies in the rejection region for the null hypothesis. The developer should reject the null hypothesis and should conclude that, based on this study, the videotape presentation is more effective for teaching basic geometric patterns to first-grade pupils than is the conventional flash-card method.

We have seen that the process of determining the t ratio from the data in the samples is exactly the same for one- and two-tail tests. The difference between the two types of tests lies in the determination of the critical value for rejection of the null hypothesis.

Let's assume that the developer in Example 15-1 had made a nondirectional hypothesis requiring a two-tail test. The hypotheses would be:

$$H_0: \mu_1 = \mu_2$$
$$H_1: \mu_1 \neq \mu_2$$

The rejection region for the null hypothesis would then be divided between both tails of the sampling distribution. For $df = 24$ at $P = .05$, Table 2 gives this critical t value as 2.064. Therefore the rejection region lies to the left of $t = -2.064$ and to the right of $t = 2.064$. The obtained $t = 2.03$ in Example 15-1, then, would not have led to rejection of the null hypothesis had a two-tail test been made.

Note that, in the same sampling distribution, we need a *t* ratio as large as 2.064 (either positive or negative) for a two-tail test, but for a one-tail test we need a *t* ratio of only 1.711. This leads us to the obvious conclusion that to be significant, an observed *t* ratio does not have to be as large in a one-tail test as it does in a two-tail test.

TESTING THE SIGNIFICANCE OF A CORRELATION COEFFICIENT

In Chapter 11 we introduced the Pearson product-moment correlation as a method of determining the degree of relationship between two variables. The statistician is generally not content with just calculating the value of the correlation coefficient. The more important problem is determining the probability that the sample coefficient could have been obtained by chance from a population where the two variables are not at all correlated. Formally stated, the statistician wants to test the null hypothesis that the correlation between the variables in the population is zero. To determine the significance of a correlation coefficient, we use Formula 15 because the distribution of sample *r*s is a function of the appropriate *t* distribution with $N - 2$ degrees of freedom.

FORMULA 15

Testing the significance of the correlation coefficient.

$$t = r\sqrt{\frac{N - 2}{1 - r^2}}$$

$df = N - 2$, where N is the number of pairs of scores

The probability that an obtained correlation coefficient could have occurred by chance in a sample drawn from a population where the correlation is zero can be determined by using Table 2 with $N - 2$ degrees of freedom.

Applying Formula 15 to the sample correlation coefficient of $r = .679$ we calculated in the example in Table 11-1, we obtain

$$t = .679\sqrt{\frac{12 - 2}{1 - (.679)^2}} = 2.925$$

We must now evaluate this *t* ratio for significance. The hypothesis being tested here is a directional one (the teacher hypothesized a positive relationship between the variables) which requires a one-tail test.

Table 2 indicates that, for degrees of freedom = 10, a *t* ratio must be at least 1.812 to be "statistically significant" at $P = .05$ (which is the level set by the teacher for significance). Because the obtained *t* ratio of 2.925 is larger than the critical value given in the table, the teacher should conclude that there is a significant positive correlation between the two variables.

We must be aware of the distinction between statistical significance and the practical utility of a correlation coefficient. As we have observed in Table 4, a small coefficient can be statistically significant when the coefficient is computed using data obtained from a very large sample. For example, in the case of a directional hypothesis, for $df = 100$, a coefficient as small as $r = .164$ is significant at the .05 level. But the coefficient of determination $r^2 = .027$ indicates that, for all practical purposes, the relationship between them is too small to be useful, since only about 3% of the variance in one variable is associated with the variance in the other variable.

Correlational hypotheses are considered nondirectional when they do not state whether the relationship is positive or negative. Such hypotheses, which simply state that there *is* a relationship, require a two-tail test of significance. Where a correlational hypothesis asserts that a relationship between two variables is positive (or that it is negative), a one-tail test is required.

DIRECTIONAL VERSUS NONDIRECTIONAL HYPOTHESES

Since most research projects are undertaken for the express purpose of demonstrating that the null hypothesis is false and should be rejected, why not always make a directional hypothesis and capitalize on the smaller critical value required to designate the rejection region? The decision to make a one- or a two-tail test should not be a capricious choice of the researcher or a matter of statistical convenience. On the contrary, the choice between the two types of tests is made not on the basis of statistical considerations but on the basis of the decisions that will be made as a result of the findings.

A nondirectional hypothesis should be made in any situation in which a finding in either direction is meaningful for decision making. For example, when the selection of one of two instructional methods is based on the outcome of a research project, findings in favor of either method dictate the adoption of that method. Here a nondirectional hypothesis must be made and a two-tail test conducted.

On the other hand, if a decision to adopt a new method is to be made only if the new method is shown to be superior to the old method, a directional hypothesis is appropriate and a one-tail test will provide the basis for rejecting the null hypothesis. Here the researcher is not concerned with the degree to which the new method might be superior to the old one; he or she is only concerned with whether it is sufficiently superior to warrant its adoption.

Directional hypotheses are sometimes made in research studies conducted to aid in development of a theory; in these cases, only one conclusion can be considered consistent with or supportive of the theory. The rationale for such directional hypotheses is that findings in the opposite direction are contraindicated by all that is known about the variables under study. This reasoning may be faulty, however, and a finding opposite in direction to the predicted outcome may be of immense value in causing modification of heretofore accepted theories.

In fact, some practitioners insist that tests in the behavioral sciences should always be two-tail. They assert that so little is known about relationships among socio-psychological variables that it is unwarranted to make hypothetical presumptions of the meaningfulness of findings in only one direction.

A safe position may be that directional hypotheses should be made only in studies where there is no useful difference, in terms of decision making, between a finding that the null hypothesis cannot be rejected and a finding that a difference exists in the direction opposite to the one predicted in the alternative hypothesis, regardless of the magnitude of that difference.

THE LEVEL OF SIGNIFICANCE

Throughout this book we have discussed examples in which either .05 or .01 was designated as the level for significance. Although these are the two conventionally adopted levels, they are by no means universal. We must realize that the researcher's choice of a level of significance prior to the experiment is somewhat arbitrary. It is merely an assertion of the risk the experimenter is willing to run of being wrong if, on the basis of sample data, a decision is made to reject the null hypothesis. In deciding on the extent of this risk, the researcher must gauge the consequences of an incorrect decision.

In general, a research study is conducted with the intent that a conventional method is to be replaced by a new one if the new method is shown to be superior to the conventional one—that is, if the null hypothesis is rejected. Making such a change from the status quo invariably entails changing a number of commonly held assumptions and practices. Consequently the researcher wants to be relatively certain that the decision to change is the correct one. The level of significance that is chosen depends on the seriousness of the consequences if the null hypothesis is falsely rejected.

In cases where the decision to change will have far reaching effects, such as in medical research, where a new medical treatment may possibly replace a conventional one, the researcher may not be satisfied with a significance level of .01. Studies of this sort may warrant a more stringent level of significance, such as .005 or .001, before the decision to change treatments is made. The researcher wants to be sure that the probability of making an incorrect decision is very low before he or she rejects the null hypothesis.

On the other hand, if a false rejection of the null hypothesis does not carry with it extremely adverse effects, a higher probability of an incorrect decision may be tolerated—for example, .10. This may be the case when the researcher is choosing between two new approaches or instructional methods.

It is considered good research procedure to set the level of significance before conducting the study. If the significance level were set after the data were collected, the researcher could be accused of choosing a level that "fits" the outcome of the experiment to some personal desires. This, of course, would be an unprofessional approach to research.

In practice, a researcher usually reports the actual significance level of the research findings if it is smaller than the preset level. For example, suppose we set .05 as the level of significance, conduct a nondirectional t test between independent sample means, and obtain $t = 3.55$. If there are 20 degrees of freedom, this finding is considered significant and the null hypothesis is rejected. We therefore report this finding as "significant beyond the .05 level" or "significant at $P < .05$." If we have an extensive table of the probability values for this particular t distribution, we can specify the actual probability level of this finding. In this example, the actual significance level is $P = .002$.

TYPE I AND TYPE II ERRORS

The purpose of conducting an experiment or research project is to provide a statistical basis on which we can make one of two possible decisions regarding the null hypothesis: to reject it or not to reject it. We have seen that neither decision can be made with certainty, because when we work with sample data a degree of sampling error is always present. To make provision for this, we set a probability level representing the amount of risk we are willing to run of being wrong if our decision is to reject the null hypothesis. Incorrect rejection of the null hypothesis is termed a *Type I error*. Another type of error can also be made. If the statistical test leads us to not reject the null hypothesis when it is in fact false, we have made a *Type II error*.

> **TYPE I ERROR**
>
> Rejecting the null hypothesis on the basis of sample data when, in fact, no difference exists.
>
> **TYPE II ERROR**
>
> Not rejecting the null hypothesis on the basis of sample data when, in fact, a true difference exists.

The choice between the two alternatives available to us and the types of possible errors can be summarized as follows:

		Null hypothesis is, in fact, TRUE	Null hypothesis is, in fact, FALSE
Our decision based on sample data	REJECT	Type I Error Probability: α	Correct Decision Probability: $1 - \beta$ (Power)
	NOT REJECT	Correct Decision Probability: $1 - \alpha$	Type II Error Probability: β

This chart indicates that there are four possible situations in connection with our decision: two involve correct decisions and two involve incorrect ones.

We have seen that the risk of a Type I error is set by the researcher prior to the experiment and is called the significance level. This probability is sometimes called the *alpha level* and is symbolized by α. Thus α represents the probability of rejecting the null hypothesis when it is true.

Let's examine more closely the Type II error. The probability of this type of error is usually called the *beta level* and is symbolized by β. Whereas the level of a Type I error is always specified by the level of significance, the Type II error, unfortunately, is sometimes neglected. Beta is the probability of not rejecting the null hypothesis when it is false. Its complement is the probability of rejecting a false null hypothesis. The probability of correctly rejecting a false null hypothesis is called the *power of the test* and is represented by $1 - \beta$.

The following questions summarize our concerns about Type I and Type II errors.

1. What probability of being wrong are we permitting ourselves if we decide to reject the null hypothesis? This is determined by the significance level we set before conducting the research study. This is the probability at which we are willing to risk making a Type I error; it is called the alpha (α) level.
2. What is the probability of our being wrong if we decide not to reject the null hypothesis? In other words, what is the probability that by not rejecting the null hypothesis we have made an error, since some alternative to it is in fact true? This is the probability of making a Type II error and is called the beta (β) level. This probability varies depending on which alternative hypothesis is involved.
3. If the null hypothesis is actually false, what is the probability that our data will lead us to the correct decision—that is, the rejection of the null hypothesis? This probability is the power of the test, and is given by $1 - \beta$.

THE POWER OF STATISTICAL TESTS

Let us examine the concept of the power of a statistical test in the context of Example 15-2.

EXAMPLE 15-2 A nondirectional hypothesis has been made that there is a difference in the effectiveness of Method A and Method B, which is revealed in achievement test scores. We set .05 as the level of significance, randomly select two groups of students, conduct the experiment, and obtain the following results.

Method A	Method B	
$N_1 = 50$	$N_2 = 60$	
$\bar{X}_1 = 70$	$\bar{X}_2 = 68.5$	$t = 1.50$
$\Sigma x_1^2 = 1,675.6$	$\Sigma x_2^2 = 1,262$	$df = 108$
		nonsignificant

. . . .

In Example 15-2, the estimate of the standard error of the difference works out to be 1.00. Because there are many degrees of freedom, the sampling distribution for this test follows the normal curve very closely. For a two-tail test with $\alpha = .05$, a t ratio at or below -1.96 or at or above $+1.96$ is required for significance. The obtained t ratio of 1.50 does not permit us to reject the null hypothesis. Since we have not rejected the null hypothesis, we have not made a Type I error.

But what if there is a true difference between μ_1 and μ_2? If this is the case, we have made a Type II error. The difficulty in determining the probability of making a Type II error (β), and thereby specifying the power of the test $(1 - \beta)$, is that in order to determine it, we must state an alternative hypothesis assigning the population parameter a specific value. Suppose two points is the smallest difference between the μ_1 and μ_2 that we are interested in detecting, if a difference exists at all. That is, if we incorrectly fail to reject the null hypothesis, a difference between μ_1 and μ_2 of less than two points is of no concern to us. To determine the power of our statistical test to lead to the rejection of the null hypothesis if $\mu_1 - \mu_2$ is at least two points, we may state the alternative hypothesis as $\mu_1 - \mu_2 = 2$. If this alternative hypothesis is true, what is the probability that we will make a Type II error? Stated more precisely, our question is: "What is the probability that the null hypothesis will not be rejected when, in fact, $\mu_1 - \mu_2 = 2$?" This probability will be β.

Figure 15-2 presents two sampling distributions: the solid-line curve represents the sampling distribution if the null hypothesis ($\mu_1 - \mu_2 = 0$) is true; the dotted-line curve represents the sampling distribution if the alternative hypothesis ($\mu_1 - \mu_2 = 2$) is true. Note that the "do not reject" region for the null hypothesis ranges from $t = -1.96$ to $t = 1.96$. Now inspect the dotted curve,

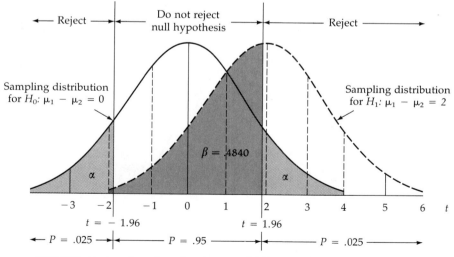

FIGURE 15-2. Sampling distribution of differences between means based on H_0: $\mu_1 - \mu_2 = 0$ and H_1: $\mu_1 - \mu_2 = 2$ from data in Example 15-2.

which has its mean centered on 2. A large proportion of this curve lies in the "do not reject" region. This area represents the probability that the null hypothesis will be incorrectly "not rejected." This is the probability of our making a Type II error, which, in this instance, is $\beta = .4840$. The power of the test is given by $1 - \beta$, which is $1 - .4840 = .5160$. This is the probability that we will correctly reject the null hypothesis if, in fact, $\mu_1 - \mu_2 = 2$. Of course, this is a very low degree of power. It is only slightly better than flipping a coin to decide whether or not to reject the null hypothesis!

We could have postulated any value for $\mu_1 - \mu_2$ and determined the power in a similar fashion. For example, if we had postulated that $\mu_1 - \mu_2 = 3$, the dotted curve in Figure 15-2 would shift to the left, so that its mean was at 3 and much less of the area under it would lie to the left of $t = 1.96$. In fact, the probability of a Type II error would be $\beta = .1492$, and the power of the test to lead to a correct rejection of the null hypothesis would be .8508.

As these two calculations demonstrate, the power of the test to lead to a correct rejection of the null hypothesis increases as the postulated difference between μ_1 and μ_2 increases. It is possible to compute the power of the test for a multitude of postulated parameter differences. Figure 15-3 presents the power function of the test of the null hypothesis based on the data in Example 15-2. The range of postulated $\mu_1 - \mu_2$ values is given along the horizontal axis, and the probability of correctly rejecting the null hypothesis (power) is given along the vertical axis. At $\mu_1 - \mu_2 = 0$, the null hypothesis is true, and the .05 probability indicated there is the probability of making a Type I error; this was the level of significance (α) that we set prior to the experiment.

Note that the curve in Figure 15-2 is based on $s_{\bar{X}_1 - \bar{X}_2} = 1.00$. Other things being equal, the larger the sample, the smaller $s_{\bar{X}_1 - \bar{X}_2}$ will be and the more

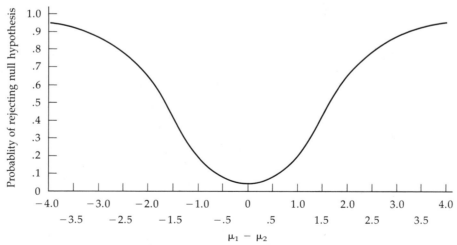

FIGURE 15-3. Power function for test of H_0: $\mu_1 - \mu_2 = 0$ for data in Example 15-2.

powerful the test that leads us to reject the null hypothesis will be. Also, other things being equal, if we reduce the probability of a Type I error, we increase the probability of a Type II error. A power function curve can be constructed for directional as well as nondirectional hypothesis tests. Again, other things being equal, the probability of making a Type II error is less for a directional than for a nondirectional hypothesis test.

In conducting an experiment, the researcher naturally wants the probability of making either type of error to be small. The researcher is directly responsible for setting the significance level, specifying its value in advance of the project. The researcher selects the probability depending on an assessment of the consequences of making a Type I error.

The power function of a statistical test can be developed and the probability of making a Type II error can be determined if the size of the smallest departure from the null hypothesis that the researcher wants to detect is specified. The β level is the probability of the risk the researcher is willing to take of not detecting a departure of that size.

In Example 15-2, we can decide to reject the null hypothesis if μ_1 and μ_2 differed by as much as 2.5 points. We can then use the power function curve in Figure 15-3 to determine that the power of this test to lead to such a rejection is .71. On the other hand, if we decide that it is important for the test to detect a difference of as much as 1.5 points, the power of the test to do so is only .32. Of course, the power of the test can be increased by increasing the size of the samples in the study.

The setting of the significance level and the evaluation of the beta level are substantive considerations of the researcher; they depend on his or her assessment of the consequences of either possible wrong decision.

EXERCISES

1. A group of social scientists formulated the following research hypothesis: "The grade point averages of students living at home are higher than the grade point averages of students living in the university dormitory." They randomly selected two samples and obtained the following data. They set $P = .05$ as the level of significance. Should the null hypothesis be rejected?

Living at home	Living in dormitory	
$N_1 = 65$	$N_2 = 70$	
$\bar{X}_1 = 3.4$	$\bar{X}_2 = 3.1$	$t = -1.30$
$s_1^2 = 1.7$	$s_2^2 = 1.8$	$df = 133$

2. In Exercise 1, if $\mu_1 - \mu_2 = .4$, what is the probability that the social scientists will make a Type II error? What is the power of this statistical test to detect that large a difference?

3. In Exercise 1, if $\mu_1 - \mu_2 = .2$, what is the probability that the social scientists will make a Type II error? What is the power of this statistical test to detect a difference that large?

4. A physical education instructor wanted to test the hypothesis that athletes who eat Diet A breakfasts will perform differently from those who eat Diet B breakfasts. He randomly assigned athletes to the two breakfast diets and, after a 6-week period, measured their athletic ability. He set $P = .01$ as his level of significance and obtained the following data. Should he reject the null hypothesis?

Diet A	Diet B	
$N_1 = 50$	$N_2 = 70$	
$\bar{X}_1 = 84$	$\bar{X}_2 = 80$	$t = 2.11$
$s_1^2 = 107$	$s_2^2 = 104$	$df = 118$

5. In Exercise 4, if $\mu_1 - \mu_2 = 9$, what is the probability that the physical education instructor will make a Type II error? What is the power of this statistical test to detect a difference that large?

6. In Exercise 4, if $\mu_1 - \mu_2 = 1$, what is the probability that the physical education instructor will make a Type II error? What is the power of this statistical test to detect that large a difference?

7. Exercise 1 of Chapter 11 presented the mechanical comprehension and divergent thinking scores for a group of students. Using $P = .05$ as the level of significance, should the null hypothesis of no correlation be rejected?

ANALYSIS OF VARIANCE—ONE WAY

Research studies often include more than two samples. Suppose we want to compare the effects of four different teaching methods on students' arithmetic achievement. If we randomly assign students to the four methods, we can measure their arithmetic achievement at the end of the school year and obtain a mean score for each of the four samples. In this study, the null hypothesis is: "There is no difference in the effectiveness of the four teaching methods on students' arithmetic achievement." This hypothesis could be symbolized by H_0: $\mu_A = \mu_B = \mu_C = \mu_D$. The alternate hypothesis is that at least one of the population means is different from the others. Note that the alternate hypothesis does not state that all population means are different; it says only that at least one of them differs from the others. Also, the alternate hypothesis does not state which of the population means will be larger (or smaller). Thus this is a nondirectional hypothesis.

To test the preceding hypothesis, we use a technique called *one-way analysis of variance for independent samples*. It is referred to as "one-way" because it tests hypotheses involving one independent variable, in this example, the method of instruction. If we can assume that the samples were randomly selected and came from normally distributed populations, the technique is appropriate. Later in this chapter, we will examine a similar technique for use with dependent samples. In Chapter 17 we will present the two-way analysis of variance technique which is appropriate for studies involving two independent variables.

If the null hypothesis is true, the population means are identical. However, we would not expect the means of our four samples to be the same; we expect some variability among them due to sampling error. We wish to ask: "If the null hypothesis is true, is the variability among the means of the samples larger than we would expect to occur by chance?" If we can show that the probability is low that the sample means differ as much as they do because of sampling error, we will reject the null hypothesis and conclude that some factor in addition to sampling error is contributing to their variation.

The ANOVA (*analysis of variance*) technique compares two estimates of the population variance to determine the probability that the difference between them is due to sampling error. One of these estimates is obtained by computing a sum of squares for each of the samples separately, and then combining these in order to obtain one population variance estimate. This is called the *within-groups variance estimate* because it is obtained by estimating variances within the samples.

The other estimate is computed by obtaining the mean score of each of the samples and then calculating a variance estimate using these mean scores and the sizes of the samples in the computation. This is called the *between-groups variance estimate* because it takes into account the variability of the means of the various samples.

Thus, through statistical procedures that will be explained later in this chapter, we obtain two estimates of the common population variance. We want to determine whether the between-groups variance estimate is significantly larger than the within-groups variance estimate. This is the crux of the ANOVA technique, because if we can show that the variance estimate based on the variability of the sample means is significantly larger than the variance estimate derived from the variability of scores within the samples, we can conclude that the samples did not come from populations having identical means. This conclusion leads us to reject the null hypothesis. In other words, if the between-groups variance estimate is too large to have occurred as a result of sampling error, we conclude that there is a difference in the effectiveness of the four teaching methods.

If we cannot reject the null hypothesis, we conclude that the samples may have come from populations with identical means and that the variation in the sample means may be due solely to sampling error. In this case, we decide that there is probably no difference in the effectiveness of the four methods of teaching arithmetic.

To apply the ANOVA technique, a ratio is computed between the two variance estimates, using the between-groups variance estimate as the numerator and the within-groups variance estimate as the denominator. This ratio is termed the *F ratio*.

In ANOVA terms, the estimate of the population variance is called a *mean square*, symbolized by *MS*, because it is the average of the sum of squares. (Recall that a variance estimate is calculated by dividing the sum of squares by the *df*, as Formula 8a shows.) We use the term "mean square" rather than "variance estimate" in connection with the ANOVA technique. Formula 34 gives the procedure for computing the *F* ratio in the ANOVA technique.

FORMULA 34

F test. Formula for computing the *F* ratio in the analysis of variance.

$$F = \frac{\text{Mean square between groups}}{\text{Mean square within groups}} = \frac{MS_b}{MS_w}$$

143
ANALYSIS OF VARIANCE— ONE WAY

To illustrate the application of the ANOVA technique, we will examine three experimental groups in Example 16-1. To gain insight into the underlying concepts involved in this technique follow the presentation closely.

EXAMPLE 16-1 To determine whether levels of illumination affect work production in an electronics firm, we randomly select 41 employees and randomly assign them to three experimental groups, with each group working under a different level of illumination. We select .05 as the level of significance for this study. After a three-month period, we measure the work production of each group and obtain the following data:

Group A		Group B		Group C	
20	16	23	19	25	22
19	15	23	18	24	21
18	15	21	18	24	21
17	14	20	18	23	21
17	13	20	17	23	21
16	12	19	16	22	20
		19	15	22	19
		19			
$N_A = 12$		$N_B = 15$		$N_C = 14$	
$\bar{X}_A = 16$		$\bar{X}_B = 19$		$\bar{X}_C = 22$	

• • • • • • •

Under the null hypothesis, we assume that all of the data in the three samples in Example 16-1 come from populations having identical means. Therefore, we form one frequency distribution of the data on the 41 employees and compute a total mean. In Example 16-1, the total mean for all 41 employees is $\bar{X}_T = 19.146$.

Figure 16-1 presents a graphic display of the frequency distributions for groups A, B, and C, and for all three groups combined. The locations of the means of the three groups and of the mean of the total group are indicated by the arrows. Here we can see that the sample means vary around the total mean. Our question is: "Is the probability .05 or less that we would get three sample means that vary this much, if they all came from normally distributed populations with the same means and variances?"

We will illustrate two methods for performing the ANOVA technique with these data. The first method consists of a step-by-step explanatory procedure so that we can grasp the meaning underlying the technique. The second method is an easier procedure for doing the actual computations. Both methods are algebraically equivalent, and they lead to identical results.

One important concept in ANOVA can be illustrated by considering the deviation of one individual's score from the mean of the total group and from

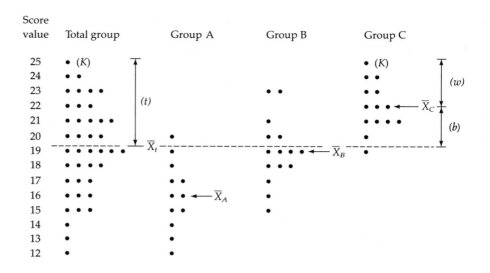

FIGURE 16-1. Frequency distributions of scores for groups A, B, and C and for all three groups combined.

his own group's mean. In Figure 16-1, subject K belongs to group C and has a score of 25. Therefore his score deviates from the total mean of 19.146 by 5.854 points, which is shown as t in Figure 16-1. Note that this deviation can be divided into two portions. One portion is w, the amount of deviation from his own group's mean, $25 - 22 = 3$ points, which is termed a deviation within a group. The other portion is b, the amount by which his group's mean deviates from the total mean, which is $22 - 19.146 = 2.854$. Thus the deviation of subject K's score from the total mean is exactly equal to its deviation from its group mean plus the deviation of the group mean from the total mean. This is true for each of the scores in all three groups. We will use these deviation scores to compute variance estimates.

Recall that the variance, as it is given in Formula 8a, is determined by dividing the sum of squares by the df. In ANOVA, the mean square (MS) is determined in the same way. Our first concern, then, is with the sum of squares. In ANOVA, the total sum of squares SS_t can be divided into two components: the sum of squares between groups SS_b and the sum of squares within groups SS_w as Formula 35 shows. These sums of squares can be calculated by applying Formulas 36 through 38 to the data in Figure 16-1.

FORMULA 35

Composition of the total sum of squares.

Total sum of squares = sum of squares between groups
+ sum of squares within groups

$$SS_t = SS_b + SS_w$$

FORMULAS 36 through 38

Calculation of sum of squares.

Total sum of squares $SS_t = \Sigma X^2 - \dfrac{(\Sigma X)^2}{N_t}$ (Formula 36)

Sum of squares within groups $SS_w = SS_A + SS_B + SS_C$ (Formula 37)

Sum of squares between groups $SS_b = SS_t - SS_w$ (Formula 38a)

$SS_b = N_A(\bar{X}_A - \bar{X}_t)^2 + N_B(\bar{X}_B - \bar{X}_t)^2 + N_C(\bar{X}_C - \bar{X}_t)^2$ (Formula 38b)

For Example 16-1, the total sum of squares and the sum of squares within each group are calculated as follows:

Using Formula 36: $SS_t = 15{,}431 - \dfrac{(785)^2}{41} = 401.12$

For each group: $SS_A = 3{,}134 - \dfrac{(192)^2}{12} = 62$

$SS_B = 5{,}485 - \dfrac{(285)^2}{15} = 70$

$SS_C = 6{,}812 - \dfrac{(308)^2}{14} = 36$

Using Formula 37: $SS_w = 62 + 70 + 36 = 168$

To determine the sum of squares between groups, we can use either Formula 38a or Formula 38b.

Using Formula 38a: $SS_b = 401.12 - 168 = 233.12$

Using Formula 38b: $SS_b = 12(16 - 19.146)^2 + 15(19 - 19.146)^2$
$+ 14(22 - 19.146)^2 = 233.12$

Of primary interest to us are the SS_w and SS_b. Examination of Formula 37 indicates that SS_w is computed by adding the sum of squares for each sample. Since the sum of squares for a particular group depends not on the magnitude of the mean score, but on the variability of scores around it, the differences in the group means do not affect SS_w; that is, the value of SS_w will be the same whether the groups all have identical means or whether their means differ greatly. (In our example, SS_w will be 168 regardless of the values of the group means.)

Formula 38b indicates that the SS_b is affected by how much each group mean differs from the total mean. Thus $SS_b = 233.12$, due to the specific differences between the group means. If the groups had identical means, SS_b would be zero. If their means differed greatly, there would be a large SS_b.

Having calculated two sums of squares, we can determine the mean squares by dividing each SS by its appropriate degrees of freedom; this procedure is shown in Formulas 39 through 41.

FORMULAS 39 through 41

Calculation of degrees of freedom for analysis of variance.

Degrees of freedom

Formula 39	$df_t = N - 1$	Total
Formula 40	$df_b =$ No. of groups $- 1$	Between groups
Formula 41	$df_w = df_t - df_b$	Within groups

In Example 16-1, $df_t = 40$, $df_b = 2$, and $df_w = 38$. Formulas 42 and 43 demonstrate how to obtain the mean squares between groups and the mean squares within groups.

FORMULAS 42 and 43

Calculation of mean squares (variance estimates).

Formula 42
$$MS_b = \frac{SS_b}{df_b}$$

Formula 43
$$MS_w = \frac{SS_w}{df_w}$$

Applying these formulas to the data in our example we obtain:

$$MS_b = \frac{233.12}{2} = 116.56$$

$$MS_w = \frac{168}{38} = 4.42$$

These mean squares represent estimates of the common population variance that are calculated from the same data but computed from two independent sources of variability, the variability within each group and the variability between the group means. Under the null hypothesis, we would expect these two estimates to be approximately the same. Obviously the MS_b in Example 16-1 is quite a bit larger than the MS_w. We now need to determine if the difference is great enough to permit us to reject the null hypothesis. Since we set our significance level at .05, we want to determine whether the probability is .05 or less that we would obtain variance estimates differing as much as these do if the null hypothesis is true. To make this decision, we must compute an F ratio using Formula 34. Using the data in our example, $F = 116.56/4.42 = 26.37$. To evaluate the significance of this F ratio, we must specify the appropriate sampling distribution of F.

THE F DISTRIBUTION

As was the case with the t distributions, there is a family of F distributions. The shape of each F distribution is determined by the degrees of freedom associated with two variance estimates. To see how an F distribution is formed, suppose we select from a normally distributed population a multitude of samples of size $N_1 = 4$, and also select a multitude of samples of size $N_2 = 31$. For each sample, we estimate the population variance s^2 using Formula 8. The df associated with the estimate derived from each sample of $N_1 = 4$ is $df_1 = 3$; the df associated with the estimate from each sample of $N_2 = 31$ is $df_2 = 30$.

If we randomly pair estimates of differing dfs and compute the ratios between the variance estimates, always placing the s_1^2 based on $df_1 = 3$ in the numerator, we obtain an array of ratios called F ratios. The ratio of each pair of estimates is determined by

$$F = \frac{s_1^2}{s_2^2} \quad \begin{array}{l} \leftarrow \text{variance estimate with } df_1 = 3 \\ \leftarrow \text{variance estimate with } df_2 = 30 \end{array}$$

The distribution of the array of F ratios calculated from all possible pairs of samples forms a sampling distribution of F. The shape of a particular F distribution depends on the df associated with the numerator and the df associated with the denominator. The particular F distribution that applies to our example, in which $df_1 = 3$ and $df_2 = 30$, is shown in Figure 16-2. As this figure shows, the lowest possible value of F ratios is zero, and this value can occur only when s_1^2 is zero. The distribution of F ratios is skewed to the right and theoretically extends to infinity. Virtually all tests of significance using an F distribution are used to determine whether s_1^2 is significantly larger than s_2^2. Therefore, when we use them, we are generally concerned with probabilities associated with areas in the right tail of the F distribution.

Figure 16-2 indicates that for this particular F distribution the probability is .05 that an F ratio will be obtained that is 2.92 or larger. The probability is .01 of obtaining an F ratio of 4.51 or larger. This represents the sampling distribution of F for 3 and 30 degrees of freedom and specifies the critical values of F commonly used in deciding to reject the null hypothesis.

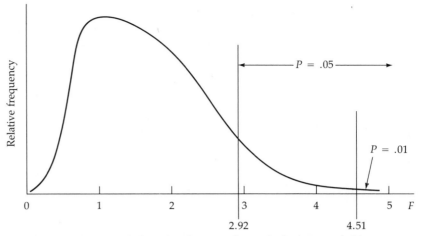

FIGURE 16-2. Sampling distribution of F with $df_1 = 3$ and $df_2 = 30$.

Since the shape of the sampling distribution of F differs for each combination of *df*s, the values of F needed at .05 and .01 differ for each combination of *df*s. Table 3 at the back of the book lists the values of F that represent the .05 and .01 areas in the right tail of each F distribution. This table is structured so that the *df*s associated with the greater mean square (variance estimate) are located across the top of the table and the *df*s associated with the smaller mean square are located along the side. Figures is roman type in the body of Table 3 indicate the F ratios needed for significance at the .05 level; figures in boldface type give the ratios needed at the .01 level.

Returning now to Example 16-1, if we want to evaluate the significance of the F ratio 26.37, we consult the columns in Table 3 showing $df_b = 2$ (for MS_b) and $df_w = 38$ (for MS_w) and find that we need an F ratio of approximately 3.23 (tabled value for the intersection of 2 and 40 *df*s) to be significant at .05. Our obtained F ratio of 26.37 is far in excess of the critical value; therefore we reject the null hypothesis that the three groups came from the same population.

The format for reporting the results of the ANOVA technique for Example 16-1 is as follows:

	Sum of squares	Degrees of freedom	Mean square	F
Between groups	$SS_b = 233.12$	$df_b = 2$	$MS_b = 116.56$	26.37 $P < .05$
Within groups	$SS_w = 168$	$df_w = 38$	$MS_w = 4.42$	
Total	$SS_t = 401.12$	$df_t = 40$		

The variance estimate based on the variability of the sample means (MS_b) has been shown to be significantly greater than the variance estimate based on within-group variability (MS_w). Therefore we conclude that some factor other than chance has affected the level of the group means, and has thereby caused the large MS_b. If our research study has been well designed, with extraneous factors controlled, we may conclude that this difference is due to the different treatments given to the three groups.

As we mentioned at the beginning of this chapter, there is an easier way to compute the sums of squares for the ANOVA technique. This method and its format are presented in Formula 44.

FORMULA 44

Computational formulas for sums of squares.

	Group A	Group B	Group C	Total
	ΣX_A	ΣX_B	ΣX_C	ΣX_t
	ΣX_A^2	ΣX_B^2	ΣX_C^2	ΣX_t^2
	N_A	N_B	N_C	N_t

Step 1 Correction term

$$C = \frac{(\Sigma X_t)^2}{N_t}$$

Step 2 Total sum of squares

$$SS_t = \Sigma X_t^2 - C$$

Step 3 Sum of squares between groups

$$SS_b = \frac{(\Sigma X_A)^2}{N_A} + \frac{(\Sigma X_B)^2}{N_B} + \frac{(\Sigma X_C)^2}{N_C} - C$$

Step 4 Sum of squares within groups

$$SS_w = SS_t - SS_b$$

Formula 44 shows that first we compute a correction term that is then used in Steps 2 and 3 to compute SS_t and SS_b. Then we find SS_w, which is the difference between SS_t and SS_b. The use of these formulas is illustrated below, using the data given in Example 16-1. Note that this simplified method yields values that are identical with those we computed using the first method.

Group A	Group B	Group C	Total
$\Sigma X_A = 192$	$\Sigma X_B = 285$	$\Sigma X_C = 308$	$\Sigma X_t = 785$
$\Sigma X_A^2 = 3{,}134$	$\Sigma X_B^2 = 5{,}485$	$\Sigma X_C^2 = 6{,}812$	$\Sigma X_t^2 = 15{,}431$
$\bar{X}_A = 16$	$\bar{X}_B = 19$	$\bar{X}_C = 22$	$\bar{X}_t = 19{,}146$
$N_A = 12$	$N_B = 15$	$N_C = 14$	$N_t = 41$

Step 1 $C = \dfrac{(785)^2}{41} = 15{,}029.88$

Step 2 $SS_t = 15{,}431 - 15{,}029.88 = 401.12$

Step 3 $SS_b = \dfrac{(192)^2}{12} + \dfrac{(285)^2}{15} + \dfrac{(308)^2}{14} - 15{,}029.88 = 233.12$

Step 4 $SS_w = 401.12 - 233.12 = 168$

Note that in the ANOVA technique, we always place the MS_b in the numerator and the MS_w in the denominator. The MS_b is almost always larger than the MS_w. Although it is possible for the MS_w to be larger than the MS_b, it is highly unlikely. This would mean that the variability of the sample means was less than would be expected by chance! (Such a finding might lead us to question whether the samples were truly randomly selected.) When an F ratio is less than 1.00, of course we cannot reject the null hypothesis. It is only when an F ratio exceeds 1.00 that we must determine whether the difference is significantly larger than we would expect as a result of sampling error. The result is that we invariably deal with only the right tail of the F distribution, even though we are testing a nondirectional hypothesis.

In studies in which the application of the ANOVA technique does not yield a significant F ratio, we must conclude that the null hypothesis of no difference among the population means is tenable. But what are we to conclude if the F ratio is large enough to permit us to reject the null hypothesis? In Example 16-1, where the null hypothesis was rejected, we can only conclude that at least one of the population means differs from the other two: the statistical test does not indicate which one differs from the others, nor does it indicate whether all three population means differ.

TESTS OF MULTIPLE COMPARISONS

One way to determine which sample or samples are significantly different from the others is to test the difference between each possible pair of means using the t test introduced earlier. However, in this case, the t test would be inappropriate because of its underlying assumption that the samples have been randomly selected. For example, if we *purposefully* select the pair of samples that have the largest difference between the means, we would be violating the assumption of randomization. Therefore, when the analysis of variance yields significant differences among the means, applying the t test to each pair of sample means is not the appropriate technique for uncovering the particular means that produced the significant finding. There are appropriate methods to use when we obtain statistical significance through using the analysis of variance technique.

Statistical tests that permit us to make comparisons among the means of various samples that have been shown to be significantly different, using the analysis of variance technique, are called tests of multiple comparisons. We present two such methods: the Tukey method for equal-sized samples and the Scheffé method for unequal-sized samples. These tests should be used to test differences between pairs of sample means whenever the ANOVA indicates there is a significant difference among the various sample means. Of course, these tests should not be used whenever the results of the ANOVA are nonsignificant, because that finding indicates that the obtained differences among all the sample means could have occurred by chance.

THE TUKEY METHOD—EQUAL-SIZED SAMPLES

The Tukey method gives a procedure for making comparisons between all possible pairs of means in a one-way ANOVA where a significant F ratio has been obtained and the sample sizes are equal. The Tukey method uses Formula 45. The expression $\bar{X}_i - \bar{X}_j$ indicates the difference between all possible pairs of samples.

FORMULA 45

Calculation of Q for the Tukey method of multiple comparisons.

$$Q = \frac{\bar{X}_i - \bar{X}_j}{\sqrt{MS_w/N}}$$

where: N = the size of each sample

Using Formula 45, a Q is obtained for the difference between each pair of sample means. Q is then evaluated for significance, using Table 4, which presents the values of Q that are required to be significant at the .05 and .01 levels for various numbers of samples based on the degrees of freedom for the within-groups mean square. The evaluation of each computed Q indicates whether the difference between a pair of sample means is considered significant.

To illustrate the Tukey procedure, consider Example 16-2, where there are equal-sized samples.

EXAMPLE 16-2 A laboratory technician supplied different types of plant food to four samples of roses to see if there was a difference in growth rates among them. She set .01 as her level of significance. After a period of plant growth, she obtained the following data.

Sample 1		Sample 2		Sample 3		Sample 4	
8	9	17	10	2	4	20	15
9	4	16	11	2	4	17	17
7	10	14	12	5	9	16	16
10	11	9	7	6	1	21	15
12	8	8	8	8	4	22	14
$\bar{X}_1 = 8.8$		$\bar{X}_2 = 11.2$		$\bar{X}_3 = 4.6$		$\bar{X}_4 = 17.3$	

· · · · · · · ·

To analyze these data, the technician used the one-way ANOVA technique for independent samples and obtained the following results:

	Sum of squares	Degrees of freedom	Mean square	F	
Between groups	844.275	3	281.425	36.222	$P < .001$
Within groups	279.700	36	7.769		
Total	1123.975	39			

Because the F ratio of 36.222 is significant, she wished to perform the Tukey procedure to ascertain which samples were significantly different from the others. Sample 1 versus sample 2 was analyzed, using Formula 45, as follows:

$$Q = \frac{8.8 - 11.2}{\sqrt{7.769/10}} = 2.723$$

Using Table 4, for within-groups degrees of freedom of 36, for four samples, at $P = .01$, a Q of at least 4.80 is needed for significance. (We used the tabled value for 30 degrees of freedom.) Thus the Q for sample 1 versus sample 2 is nonsignificant. The Q values for the other pairs of samples are computed and evaluated in the same way. Listing the samples in order of the magnitude of their means, and placing an asterisk where Q was significant, she found:

Sample 3 $\bar{X}_3 = 4.6$
Sample 1 $\bar{X}_1 = 8.8$
Sample 2 $\bar{X}_2 = 11.2$
Sample 4 $\bar{X}_4 = 17.3$

Sample 3 vs. Sample 1 $Q = 4.765$
Sample 3 vs. Sample 2 $Q = 7.488*$
Sample 3 vs. Sample 4 $Q = 14.408*$
Sample 1 vs. Sample 2 $Q = 2.723$
Sample 1 vs. Sample 4 $Q = 9.643*$
Sample 2 vs. Sample 4 $Q = 6.920*$

Therefore the technician concluded that between samples 1 and 2 and between samples 1 and 3 there were no significant differences, but all other pairings were significant, with sample 4 having a significantly larger mean than any of the other three samples.

THE SCHEFFÉ METHOD—UNEQUAL-SIZED SAMPLES

Where a one-way ANOVA reveals a significant F ratio between unequal-sized samples, the Scheffé method for multiple comparisons is appropriate for making comparisons between sample means. To perform this test, each possible pair of sample means is examined, using Formula 46.

FORMULA 46

Calculation of F for the Scheffé method of multiple comparisons.

$$F = \frac{(\bar{X}_i - \bar{X}_j)^2}{MS_w(1/N_i + 1/N_j)}$$

FORMULA 47

Determination of F' for use in evaluating the significance of F for the Scheffé method.

$$F' = (k - 1)F$$

where: F = the critical value from Table 3, with:
df for numerator: $k - 1$
df for denominator: $N - k$

where: k = number of samples
N = number of subjects in all samples

The F obtained for each comparison is evaluated for significance against F' as obtained using Formula 47. Note that Formula 47 requires us to determine the critical value of F from Table 3, using the appropriate degrees of freedom, and then multiplying it by $k - 1$ to obtain F'. This F' then becomes the critical value for evaluating the obtained F ratios calculated by Formula 46. All F ratios that exceed the value of F' are considered as statistically significant. Example 16-3 illustrates this technique.

EXAMPLE 16-3 A highway inspector wanted to determine if there is a difference in the asphalt durability between the products of four asphalt producers. He set $P = .05$ as his level of significance and analyzed specimens from each producer and obtained the following durability ratings.

Sample 1		Sample 2		Sample 3		Sample 4	
4	10	14	24	6	1	14	17
7	4	10	22	8	3	15	16
3	8	9	17	6	9	20	14
2	8	17	22	4	7	27	22
9	7	19	21	2	6	27	24
6	2	20		1	5		
$N_1 = 12$		$N_2 = 11$		$N_3 = 12$		$N_4 = 10$	
$\bar{X}_1 = 5.833$		$\bar{X}_2 = 17.727$		$\bar{X}_3 = 4.833$		$\bar{X}_4 = 19.600$	

· · · · · · ·

The one-way ANOVA technique applied to the data in Example 16-3 yields the following:

	Sum of squares	Degrees of freedom	Mean square	F	
Between groups	2001.285	3	667.095	42.476	$P < .001$
Within groups	643.915	41	15.705		
Total	2645.200	44			

Because the F ratio is statistically significant, it is proper to perform a test of multiple comparisons. The samples being of different sizes, Scheffé's test is the appropriate method to use.

Applying Formula 46 to the means of samples 1 and 2, we obtain:

$$F = \frac{(5.833 - 17.727)^2}{15.705(1/12 + 1/11)} = 51.696$$

Performing similar calculations for each pair of samples we have:

$$\begin{array}{ll} \text{For } \bar{X}_1 \text{ vs. } \bar{X}_2 & F = 51.696* \\ \text{For } \bar{X}_1 \text{ vs. } \bar{X}_3 & F = .382 \\ \text{For } \bar{X}_1 \text{ vs. } \bar{X}_4 & F = 65.822* \\ \text{For } \bar{X}_2 \text{ vs. } \bar{X}_3 & F = 60.754* \\ \text{For } \bar{X}_2 \text{ vs. } \bar{X}_4 & F = 1.170 \\ \text{For } \bar{X}_3 \text{ vs. } \bar{X}_4 & F = 75.732* \end{array}$$

To evaluate the significance of each of these F ratios, we must first use Formula 47 to determine the critical value of F'. Recall that the inspector set $P = .05$ as his level of significance. Using Table 3, with 3 and 44 degrees of freedom, we find that the critical value of F is 2.84 (the closest tabled value being for 3 and 40 degrees of freedom). Applying Formula 47, we obtain:

$$F' = 3(2.84) = 8.52$$

Thus F ratios that are 8.52 or larger are considered significant. In the preceding list, these F ratios are asterisked. The inspector concludes from this information that samples 1 and 3 do not differ, that neither do samples 2 and 4, but that samples 2 and 4 are both more durable than either sample 1 or 3.

EXERCISES

1. A reading specialist wanted to determine if there was a difference in the effectiveness of three methods of remedial reading instruction. He randomly assigned students to the Individual Method, Small Group Method, and Large Group Method and tested them at the conclusion of the instructional period on reading speed and obtained the following reading scores. He set $P = .05$ as the level of significance. Should the null hypothesis be rejected? If so, which method(s) can be considered as significantly different from the others?

Individual	Small group	Large group
30	32	40
29	31	38
28	31	37
27	30	37
27	30	36
27	30	36
24	27	
21		

2. Use the data presented in Exercise 1 of Chapter 11. Should the industrial arts teacher reject the null hypothesis that there is no relationship between

the mechanical comprehension and divergent thinking scores of students? Use $P = .01$ as the level of significance.

3. Four samples of white rats were randomly selected and given different types of liquid diets for a two-week period. At the end of the period, the following weights were recorded. Using $P = .01$ as the level of significance, should the null hypothesis that the liquids have no differential effects on the weights of the white rats be rejected? If so, which diet(s) can be considered as significantly different from the others?

Diet 1	Diet 2	Diet 3	Diet 4
10	11	19	10
11	12	23	11
13	13	23	12
13	13	25	13
13	15	27	14
15	9	26	16

4. A dean of education wanted to test the hypothesis that there is no effect of instructional styles on the concept acquisition of college seniors in a curriculum methods course. She randomly assigned students to three sections of the same course, each taught by a professor having a different instructional style and obtained the following scores on a concept final examination. She set $P = .05$ as the level of significance. Should she reject the null hypothesis? If so, which instructional style(s) is significantly different from the others?

Lecture only		Lecture-participation		Participation only	
35	31	40	38	37	38
31	29	38	36	36	36
34	32	37	41	34	30
33	32	42	35	40	38
32	27	37	37	35	33
30	35	39	39	32	36
33	34	40	33	36	35
28	33	38	41	39	40
37	29	45	37	35	34
32	31	36	43	37	32
29	36	37	39	36	37
31	32	42	35	34	35
33	30	39	39	38	36
30	35	40	36		
36	32	39	40		
32	29				
35	33				
33	34				
31					

5. An automobile manufacturer decided to test four brands of seat covers before selecting one of them for installation on new model cars. He ordered thirty of each brand, tested them for wearability, and recorded the following wearability scores for the seat covers. The level of significance was set at

$P = .01$. Should the manufacturer reject the null hypothesis? If so, which brand(s) should be considered more durable than the others?

Brand 1		Brand 2		Brand 3		Brand 4	
52	51	56	52	55	56	51	50
50	50	51	50	59	55	50	56
51	54	49	54	57	57	54	48
49	50	54	49	56	55	47	51
54	52	50	51	54	60	51	46
50	47	57	50	56	54	49	50
48	51	48	50	55	57	52	44
55	48	53	52	60	59	49	50
49	53	51	50	56	59	48	49
53	49	45	56	54	56	52	54
47	55	49	48	58	55	47	50
50	51	53	55	52	56	48	49
45	50	51	50	58	55	52	51
52	52	49	48	51	58	50	46
54	50	55	53	57	56	46	51

17 TWO-WAY ANALYSIS OF VARIANCE AND OTHER TECHNIQUES

Up to this point, we have accompanied the presentation of each statistical technique with formulas and an illustration of exactly how the calculations are performed. This was to familiarize you with the way in which the computer programs function in processing data, and to give you some "hands on" experience with data analysis. Of course, the computer programs are available to perform the computational chores for you and are certainly an asset because of their speed and accuracy. Owing to the computational complexity involved in performing the three techniques presented in this chapter, we will not deal with the many new symbols, some involving multiple subscripts, that would be necessary to do the calculations by hand. Instead, we will focus on the conceptual understanding of the technique. To actually do the computations necessary to apply these techniques, you should seek out a computer program.

TWO-WAY ANALYSIS OF VARIANCE

In Chapter 16 we considered a method of data analysis, the one-way ANOVA, that involved one independent variable and one dependent variable. The logic of this approach can be extended to permit the analysis of more variables in a single analysis. A useful technique, called two-way analysis of variance, permits the simultaneous examination of the effects of two independent variables on a dependent variable. To illustrate this technique, consider the problem posed in the following example.

EXAMPLE 17-1 A pharmaceutical company developed three new drugs to relieve pain. Because of the particular ingredients in these drugs, the company was asked to conduct research on the effect of the drugs on relieving pain for individuals having differing tolerance levels for alcohol. To determine these effects the company conducted a study in which the researcher measured the effectiveness of each drug on three samples of subjects; those having low, medium, and high tolerance levels for alcohol. The researcher set $P = .01$ as the level of significance and recorded the following pain-relief scores for each drug type according to alcohol tolerance levels.

Tolerance for alcohol

	Low		Middle		High	
Drug 1	75	84	60	42	22	35
	80	85	40	57	40	18
	79	79	55	59	30	32
	92	90	50	46	37	25
	$\bar{X} = 83.000$		$\bar{X} = 51.125$		$\bar{X} = 29.875$	
Drug 2	12	18	19	15	11	7
	10	15	17	12	14	20
	19	17	10	13	14	17
	18	20	9	13	9	16
	$\bar{X} = 16.125$		$\bar{X} = 13.500$		$\bar{X} = 13.500$	
Drug 3	27	38	60	71	87	84
	32	29	70	79	90	81
	40	27	73	64	79	74
	35	25	68	62	85	91
	$\bar{X} = 31.625$		$\bar{X} = 68.375$		$\bar{X} = 83.875$	

• • • • • • •

This is an appropriate statistical task for the two-way analysis of variance technique because it involves two independent variables—the type of drug and the level of alcohol tolerance. The dependent variable is the amount of pain relief that the subjects reported after taking their assigned drug. Of course it is possible to do a one-way ANOVA on the data in Example 17-1 to determine if there was a difference in the effectiveness of the three drugs on pain relief. Had the researcher done this, the following results would have been obtained. To verify the following table for Type of Drug, you must combine all data cross the three levels of alcohol tolerance for each type of drug.

One-way analysis of variance for three types of drugs

	Sum of squares	Degrees of freedom	Mean square	F	
Between groups	30,947.861	2	15,473.931	42.555	$P < .001$
Within groups	25,089.917	69	363.622		
Total	56,037.778	71			

Based on these findings, the researcher would have concluded that there was a significant difference in the effectiveness of the three drugs, but he would have no information about how the effectiveness was related to the level of alcohol tolerance. To do this, he must use the two-way ANOVA technique, which examines both type of drug and level of alcohol tolerance simultaneously as they affect pain relief.

An examination of the sample means reveals that some differences exist as a result of level of alcohol tolerance, and these appear to differ for each type of drug. Whereas the one-way ANOVA technique was useful for testing one hypothesis regarding treatment means, the two-way ANOVA technique permits us to test three hypotheses. It allows us to:

1. Test the difference in the type-of-drug mean scores (represented by the three rows of data).
2. Test the difference in the tolerance-level mean scores (represented by the three columns of data.)
3. Examine a different type of hypothesis than has hitherto been possible with a t test or the one-way ANOVA.

Now we can examine the question: Do the drugs have different effects on pain relief as a result of different levels of alcohol tolerance? This is called a test of the interaction effect, so called because it examines the way in which drug type and alcohol tolerance level interact to affect pain reduction. This ability to examine the interaction of independent variables on a dependent variable is an important addition to our array of statistical procedures afforded us by the two-way ANOVA technique.

Although we will not illustrate the actual step-by-step process of doing the computations for the two-way ANOVA,* we will present the approach used in this technique. Example 17-1 can be referred to as a 3 × 3 ANOVA, because there are three rows and three columns which encompass the nine cells of data.

In the one-way ANOVA, recall that we partitioned the total sum of squares into two components: the between sum of squares and the within sum of squares. In the two-way ANOVA, we partition the total sum of squares into four components as follows:

$$SS_t = SS_r + SS_c + SS_{rc} + SS_w$$

where r stands for row, c stands for column, and w stands for within-cell. The symbols in the preceding equation and how the values for them are computed are as follows:

SS_t (the total sum of squares) is obtained by summing the squares of the deviation of each score in the total study from the grand mean (the mean of all the scores in the study).

---

* See Hinkle, D., Wiersma, W., and Jurs, S., *Applied Statistics for the Behavioral Sciences*. Chicago: Rand McNally College Publ. Co., 1979, pp. 296–328.

SS_r (the row sum of squares) is obtained by summing the squares of the deviation of each row mean from the grand mean, and multiplying this sum times cn (the number of columns times cell size).

SS_c (the column sum of squares) is obtained by summing the squares of the deviation of each column mean from the grand mean, and multiplying this sum times rn (the number of rows times cell size).

SS_{rc} (the interaction sum of squares) is obtained by summing the squares of the deviation of each cell mean from its predicted value, assuming that no interaction exists, and multiplying this sum times the cell size.

SS_w (the within-cells sum of squares) is obtained by summing the squares of the deviation of each score from its cell mean.

The total number of degrees of freedom $(N - 1)$ also may be partitioned into components associated with each of the sum of squares, as follows:

$$\text{total } df = \text{row } df + \text{column } df + \text{interaction } df + \text{within } df$$

where total $df = N - 1$, row $df = r - 1$, column $df = c - 1$, interaction $df = (r - 1)(c - 1)$, and within $df = \Sigma(\text{cell } N - 1)$.

To obtain the variance estimates needed for the two-way ANOVA, divide each sum of squares by its appropriate degrees of freedom to obtain the mean square (a variance estimate similar to those obtained for the one-way ANOVA). In effect, each mean square can be taken as an estimate of the population variance. The within-subjects mean square, being unaffected by any differences due to type of drug or level of alcohol tolerance, is taken as the variance estimate against which the other mean squares are compared.

The between-rows, between-columns, and interaction mean squares are divided by the within-subjects mean square to obtain three F ratios (each representing one of the three hypotheses being tested). Each F ratio is then compared with the critical value of F based on the appropriate degrees of freedom and the preset level of significance to determine whether or not each null hypothesis should be rejected.

The following table presents the results of a two-way ANOVA using the data in Example 17-1.

	Sum of squares	Degrees of freedom	Mean square	F	
Between rows	30,947.861	2	15,473.931	466.160	$P < .001$
Between columns	44.778	2	22.389	.674	$P = 1.000$
Interaction	22,953.889	4	5,738.472	172.874	$P < .001$
Within cells	2091.250	63	33.194		
Total	56,037.778	71			

The F ratio of 466.10 between rows (drug type) is evaluated using $df = 2$ for the numerator and $df = 63$ (within cells) as being highly significant, whereas the between columns (alcohol tolerance level) is nonsignificant. If these were the only findings available, the researcher might be misled into concluding that the drugs differ in effectiveness but alcohol tolerance level has no effect. However, the two-way ANOVA reveals a highly significant interaction effect. To interpret what this effect is, it is necessary to examine the cell means. An effective method for doing this is to prepare a graph of the mean scores, as is presented in Figure 17-1.

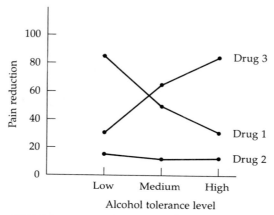

FIGURE 17-1. Mean scores for Example 17-1.

From Figure 17-1 it is evident how the two-way ANOVA findings occurred. First, Drug 2 had little effect on pain reduction regardless of the alcohol tolerance level. Second, Drug 1 is highly effective with low alcohol tolerance subjects and decreases in effectiveness with increased alcohol tolerance. The opposite occurred with Drug 3, it being more effective with high alcohol tolerance subjects. The importance of the two-way ANOVA technique is quite evident here, in that the research may well conclude that Drug 2 is ineffective regardless of alcohol tolerance level, whereas Drug 1 or Drug 3 is quite effective, depending on the alcohol tolerance level. Without the two-way ANOVA technique, we would not have been able to detect all of the relationships in one statistical analysis of data.

Constructing a graph, as we did Figure 17-1, is an excellent way to show how the various cell means lie in relation to each other. If the lines in such a graph are parallel, there is no interaction effect. However, the lines that are not parallel (whether or not they actually cross) contribute to the interaction mean square. If the lines are sufficiently nonparallel, a significant interaction effect occurs.

As was the case with the one-way ANOVA, multiple comparisons may be made when there is a significant difference among the rows, or among the columns, or both. The Tukey test of multiple comparisons may be used to compare the row means (which, in Example 17-1, represent the three drugs) because they were shown to be significantly different in the ANOVA. This test calculates Q for each of the possible pairs of row means. Each Q is evaluated for significance using Table 3. This same process could be applied to the column means (the three levels of alcohol tolerance) had the F ratio for columns been significant.

Because the different types of drugs produced a significant difference in Example 17-1, the Tukey test, applied to the rows, gives the following results.

$$\text{Row 1 vs. Row 2} \quad Q = 34.260$$
$$\text{Row 1 vs. Row 3} \quad Q = 5.633$$
$$\text{Row 2 vs. Row 3} \quad Q = 39.893$$

Each of the preceding Q values is evaluated for significance, using Table 4 to determine the critical Q for three groups with $df = 63$.

Table 4 indicates that a critical Q of 3.40 (table value at $df = 60$) or larger is needed to reject the null hypothesis at the .05 level of significance. Each of the obtained Q values is significant. Therefore the researcher should conclude that a difference exists between each pair of drug types under consideration.

This technique can be used only if the following are true:

1. The samples are of equal size.
2. The samples are randomly selected.
3. The variances of the populations from which the samples have been selected are equal.
4. The distribution of values on the dependent variable in the population from which the samples are selected is normal.

There are extensions of this technique to three or more independent variables (and even more than one dependent variable), but these are beyond the scope of this text.*

ANALYSIS OF COVARIANCE

Researchers always want to reach valid conclusions regarding the results of their studies. To do this, they go to great lengths to ensure that if they obtain a difference on the dependent variable, this difference is due to the independent variable or variables under study, and not due to some other extraneous variables. Direct control or statistical control may be employed to control the effects of such extraneous variables.

The preferred method is *direct control* of the variable. Usually this control is achieved either by randomly assigning subjects to the various groups, which ensures (within sampling error) that the samples are equated on all variables, or by matching subjects on a crucial variable, which ensures that they are equated on that variable. However, these techniques are not always possible,

---

* See Kirk, R. E., *Experimental Design: Procedures for the Behavioral Sciences, Second Edition*. Monterey, CA: Brooks/Cole, 1982.

and the researcher must resort to *statistical control* to neutralize the effect of an extraneous variable. This type of control is achieved by using the analysis of covariance (ANCOVA) technique. It is particularly useful whenever intact groups are used as samples and it is known (or suspected) that the groups differ on a variable that is related to the dependent variable in the study.

In the analysis of covariance technique, the variable we want to control statistically is called the covariate. The function of this technique is to remove or neutralize, by statistical procedures, the effect that the covariate has on the dependent variable when studying the effect of the independent variable. To illustrate this technique, consider Example 17-2.

EXAMPLE 17-2 A high school principal wants to try out four different curricular approaches to teaching geometry to tenth-grade students in his school. He selects four existing geometry classes and asks each teacher to teach a different approach for one school year. Because of the method of assigning students to classes, he suspects that the aptitude levels of the students are different across the four classes. He cannot reassign students for the purpose of testing the different approaches. However, he does have the scholastic aptitude scores of the students on file. At the end of the school year he obtains the scores on the final geometry examination given to all classes. He sets $P = .05$ as the level of significance. The students' aptitude scores and the geometry final examination scores are as follows.

Group 1		Group 2	
Aptitude	Geometry	Aptitude	Geometry
27	81	17	72
29	85	16	74
33	78	19	77
24	79	22	78
31	87	24	78
32	88	20	68
28	68	19	64
29	62	16	52
61	84	15	60.
32	81	14	71

Group 3		Group 4	
Aptitude	Geometry	Aptitude	Geometry
12	66	32	84
9	62	31	83
14	63	27	74
17	69	24	77
12	59	20	63
10	42	25	64
17	68	19	65
16	61	23	81
24	77	20	89
10	63	24	77

. . . .

Before we examine the use of the ANCOVA technique with the data in Example 17-2, let's suppose the principal did not take into account the scholastic aptitude scores and just performed a one-way ANOVA on the geometry scores. A one-way analysis of variance applied to these data would yield an F ratio of 6.732, which is significant at $P < .001$. Therefore, on the basis of this test, the null hypothesis of no difference is rejected and the principal would conclude that there was a significant difference among the four curricular methods.

Based on this finding, the principal may be misled into assuming that the difference in the groups' means was *caused* by the different curricular methods used. However, we must remember that these are intact groups and the scholastic aptitude of the students may have some differential effect on the geometry scores. To begin our examination of this, we should compute a Pearson correlation (r), which gives a measure of the relationship between aptitude and geometry scores. Combining the data across the four groups, for an N of 40, we get a correlation coefficient of $r = .635$, which is significant at $P < .001$, indicating a fairly strong relationship between the variables. Now, let's look at the mean geometry and mean aptitude scores for the four groups.

	Group 1	Group 2	Group 3	Group 4
Geometry mean	32.6	18.2	14.1	24.5
Aptitude mean	79.3	69.4	63.0	75.7

Clearly the groups with higher geometry means also have the higher aptitude means. In review, we have demonstrated by ANOVA that there is a significant difference among the geometry means for the four groups. Also, there is a correlation between geometry and aptitude scores, and the geometry and aptitude means of the four groups are in the same order. The question the principal now raises is: "What would be the geometry mean scores if the groups didn't differ in aptitude?" In other words, how can an adjustment be made that will remove the effects of the aptitude scores on the geometry scores?

Here is where the analysis of covariance technique may be used profitably. The ANCOVA technique compares the differences between the geometry mean scores when the influence of aptitude scores has been statistically controlled or removed. In effect, this technique combines both regression analysis and analysis of variance methods. In using this technique, the dependent variable (geometry scores) is called the criterion and is denoted by Y. The uncontrolled variable (aptitude scores) is called the covariate and is denoted by X. The regression method is incorporated in the technique to remove the influence of the covariate on the criterion variable. (Of course, if there is no correlation between X and Y, performing a covariance analysis would have no effect on the means of the criterion variable since there would be no relationship, hence no effect, between them to remove.)

The ANCOVA approach lends itself to more complex applications than we are presenting here, including the use of multiple covariates. In our example, which is the simplest form of ANCOVA, we are dealing with three variables:

Independent variable: Method of instruction
Dependent variable (Y):   Geometry final exam scores
Covariate (X):   Scholastic aptitude scores

The use of the ANCOVA technique yields adjusted mean scores for the dependent variable, the means being adjusted to remove the effect of the covariate on them. Then a test of significance is applied to the adjusted means. We will not present the formulas for making the actual calculations;* we leave that complex task to computers. If we had one at our disposal to compute an analysis of covariance on the data in Example 17-2, the following data would be obtained.

Analysis of covariance for data in Example 17-2

	Sum of squares	df	Mean square	F
Between treatments	197.44	3	65.813	.971 n.s.
Error	2372.99	35	67.800	
Total	2570.43	38		

	Mean of covariate	Mean of criterion	Adjusted mean of criterion
Group 1	32.600	79.300	73.971
Group 2	18.200	69.400	71.557
Group 3	14.100	63.000	67.289
Group 4	24.500	75.700	74.582

The ratio of .971 is nonsignificant. This statistical test is of the difference between the adjusted criterion means, the effects of the covariate having been statistically removed. Thus the findings indicate that, whereas a significant difference was evident in the ANOVA of the original geometry scores, when the influence of the aptitude scores is removed, the resulting difference among the geometry means is nonsignificant.

An examination of the mean scores just presented gives some insight into the effect of the ANCOVA technique. The means of the covariate (aptitude scores) and the criterion variable (geometry scores) are the same as presented earlier. The adjusted means of the groups on the criterion variable are the product of the ANCOVA technique. These means represent the projected geometry mean scores the groups would have obtained if they had had equal aptitude means. Note that group 1's mean score is adjusted downward because it had the highest aptitude mean, whereas group 3's mean is adjusted upward because it had the lowest aptitude mean score. Also, the adjusted criterion means are much less variable than the unadjusted ones, indicating that the removal of the effects of the covariate substantially reduced the difference among these means. The adjusted means are so much more homogeneous that the

---

* See Ferguson, G. A., *Statistical Analysis in Psychology and Education* (5th Ed.). New York: McGraw-Hill, 1981, pp. 358–374.

removal of the effects of the covariate resulted in a nonsignificant F in the ANCOVA test.

ONE-WAY ANALYSIS OF VARIANCE WITH REPEATED MEASURES

A common type of research design requires that repeated measurements be taken on the same subjects under different treatments or at different time points during treatment. Earlier we presented a method for analyzing the data in the simplest repeated measures design where data were obtained for one group under two conditions, or at two time points, called the *t* test for dependent samples. We now discuss a method that is appropriate for analyzing repeated measures data collected under more than two conditions or time points.

A one-way analysis of variance with repeated measures is used to determine if the correlated sets of data have statistically different mean scores. This approach takes into account the possibility that if the same subjects are measured under more than one condition, the variability of scores in the conditions may be correlated. The presence of such correlation reduces the error term, thereby allowing us to capitalize on a situation that does not exist when dealing with independent samples. The use of this program necessarily requires equal group sizes, since the data in each group are for the same individuals. Rather than present the step-by-step method of performing the necessary calculations, we will let Example 17-3 illustrate the use of this technique.*

EXAMPLE 17-3 A manufacturer of engines wanted to determine if his engines operate at different levels of efficiency under differing operating temperatures. He randomly selected twelve engines from his inventory and obtained efficiency ratings for each engine under four different temperatures. He set $P = .01$ as his level of significance. The efficiency ratings are as follows:

Temperature levels (in degrees centigrade)

	60°	70°	80°	90°
Subject 1	93	96	80	72
Subject 2	89	81	86	84
Subject 3	96	87	83	91
Subject 4	87	86	90	80
Subject 5	86	84	82	78
Subject 6	88	85	85	79
Subject 7	79	83	75	82
Subject 8	85	77	80	73
Subject 9	84	81	79	78
Subject 10	83	82	81	70
Subject 11	79	75	72	75
Subject 12	80	77	77	76

- - - - - - -

*See Ferguson, G. A., *Statistical Analysis in Psychology and Education* (5th Ed.). New York: McGraw-Hill, 1981, pp. 316–321.

The layout of the data is analogous to the two-way ANOVA, with each subject representing a row where each row has $N = 1$.

Applying a one-way ANOVA with repeated measures to the data in Example 17-3, we obtain the following results.

Analysis of variance—repeated measures for data in Example 17-3

	Sum of squares	df	Mean square	F	
Between groups	369.229	3	123.076	7.425	$P < .001$
Within subjects	736.229	11	66.930		
Interaction	547.021	33	16.576		
Total	1652.479	47			

Temperature	Mean
60°	85.750
70°	82.833
80°	80.833
90°	78.167

The format of the preceding table resembles ANOVA tables encountered previously, with the exception that the F ratio is calculated by dividing the mean square for groups by the mean square for interaction, there being no "error" term calculated by this technique. Also, an F ratio for subjects is not calculated because this technique does not permit this calculation. This is no drawback, however, since there is usually no interest in determining if subjects differ from each other.

The obtained F ratio of 7.425 is evaluated using 3 and 33 degrees of freedom and is significant at $P < .001$. Thus, by examining the mean scores, the manufacturer concludes that temperature has an inverse effect on the efficiency levels of his machines.

18 CHI-SQUARE TESTS

In the preceding chapters we explored a variety of statistical techniques for testing research hypotheses. In all of the tests presented thus far we have made several assumptions regarding the population distribution from which the sample or samples were selected. One of these assumptions was that the variable or variables in the population under study be normally distributed. Other assumptions were made for specific tests. For instance, there were different approaches to using the t test depending on whether or not the variances were assumed to be equal, and we had to assume that the variables analyzed using the Pearson product-moment correlation were linearly related. The hypotheses being tested in these situations were about population parameters. Therefore, such tests are called *parametric tests*.

However, the researcher often has little or no information about the population parameters, or the data in the sample or samples indicate that the assumptions underlying a parametric test are clearly unwarranted. Statistical methods that do not require assumptions about population parameters are termed *nonparametric* statistical techniques.

If nonparametric techniques are available, the question arises: "If nonparametric techniques do not require us to assume anything about the shape of the population distribution, why don't we always use them rather than the parametric techniques?"

Parametric techniques are preferred when they are appropriate because they are more powerful than their nonparametric counterparts. If a correlation (or difference between means) exists in the population, a parametric technique is more likely to produce a significant finding; thus the probability of making a Type II error (not rejecting a false null hypothesis) is less with a parametric technique.

Also, studies have shown that parametric tests are quite robust when the underlying assumptions are violated. This means that even if a population distribution is quite nonnormal in shape, the parametric tests still tend to give us valid findings. This is especially true when sample sizes are large.

For these reasons, we find parametric tests used even when the assumption of normality in the population is questionable. On the other hand, the use of nonparametric techniques is becoming increasingly widespread, especially when small samples are involved.

Nonparametric techniques are appropriate whenever data are on the nominal or ordinal scale; parametric techniques should be used for analyzing data on the interval and ratio scales whenever the assumptions underlying the techniques can justifiably be met. If these assumptions cannot be met, the interval or ratio data are often converted to nominal or ordinal form and then nonparametric techniques are applied.

This chapter goes into some detail on a frequently used nonparametric technique—chi square. In the next chapter we present a selected number of nonparametric techniques that are analogs to some of the parametric techniques presented thus far.

Statistical problems are frequently encountered in which the data are in the form of frequencies rather than score values; in these cases, our job is to determine whether the distribution of the frequencies across a set of categories differs from a set of expected frequencies. The statistical procedure that is appropriate for such problems is a commonly used nonparametric technique called *chi square*. It is symbolized by χ^2, using the Greek letter chi (rhymes with eye). Whenever data can be classified into a set of mutually exclusive categories, we can use a chi-square sampling distribution to determine the probability that the distribution of observed frequencies differs from the distribution of expected frequencies, based on a given hypothesis.

The chi-square technique is appropriate to use when we are solving two basic types of problems involving the comparison of observed and expected frequencies. These two types, which are examined in this chapter, are the test for goodness of fit and the test for independence. The two types differ in the hypothesis that is to be tested and in the method by which the expected frequencies are determined. Both tests involve the use of a chi-square sampling distribution.

THE TEST FOR GOODNESS OF FIT

Problems in which we want to determine whether the distribution of frequencies across a set of categories observed in a sample can reasonably be regarded as fitting a hypothetical set of frequencies in a population require a goodness-of-fit test. Example 18-1 illustrates the simplest form of a study involving the goodness-of-fit test.

EXAMPLE 18-1 A marketer wanted to determine whether there was a difference in the preference of American business executives for brand A and brand B cigarettes. He randomly selected 50 business executives, had each of them try both brands, and noted that 17 preferred brand A and 33 preferred brand B. He chose $P = .05$ as his significance level for making a statistical test.

. . . .

Note that the type of data collected in Example 18-1 is not in the form of scores or ranks, but in terms of frequencies of responses to the two brands of cigarettes. These data represent the nominal level of measurement since they are merely frequencies that fall into alternative categories. In addition, the data represent frequencies of occurrence of discrete events; thus they represent discrete measurements rather than continuous variables. This means that an individual event, such as the choice between brands in our example, falls either into one category or into the other; a person's choice cannot be split between categories. The result is that the data are in the form of integers, such as 40–10, 38–12, or 29–21, indicating that one brand is favored over the other.

Using the chi-square technique, we can determine the probability that the frequencies we observe in our sample differ from a set of hypothesized frequencies. In Example 18-1, the null hypothesis being tested is that there is no difference between the number of business executives in the population who prefer brand A and the number who prefer brand B. If the null hypothesis is correct, we would expect 25 of the 50 executives to choose brand A and 25 to choose brand B. In the chi-square test, these are called the expected frequencies; that is, they are the frequencies we would expect to occur by chance.

Since we found that 17 business executives preferred brand A and 33 preferred brand B, these are called the observed frequencies. The data indicate that more executives preferred brand B than brand A, but, as is true of all situations in which we obtain data from samples, there is a possibility that the difference between the preferences is due to sampling error and not to a true difference in the population. Our question, then, is: "Are the observed frequencies sufficiently different from the expected frequencies to justify rejection of the null hypothesis?"

The chi-square test provides us with a statistic based on the differences between observed and expected frequencies. The test tells us at what level of probability (for instance, the $P = .01$ or the $P = .05$ level) the difference between observed and expected frequencies is significant. Thus, by this test, we determine whether the observed frequencies in our sample differ significantly from the expected frequencies based on the null hypothesis. If they do, we reject the null hypothesis and conclude that the population of business executives prefers brand B over brand A. If they do not differ significantly, we conclude that the difference in frequencies obtained from our sample may be due to sampling error.

In preparation for chi-square analysis, the observed and expected frequencies in Example 18-1 are presented in the following summary table:

	Observed (O)	Expected (E)
Brand A	17	25
Brand B	33	25
Total	50	50

We are now ready to test the null hypothesis using the chi-square technique. Formula 49 is the formula for the simplest type of chi-square analysis, the type used in Example 18-1, where observed frequencies fall into two categories, which are called *cells* in chi-square tests.

FORMULA 49

Calculation of the chi square when $df = 1$.

$$\chi^2 = \Sigma \frac{(|O - E| - .5)^2}{E}$$

where: O = observed frequency
  E = expected frequency

The subtraction of .5 from each $|O - E|$ represents Yates' correction for continuity.

In using Formula 49 to compute χ^2, we are concerned with the difference between O and E in each cell. In the formula the two vertical bars encompassing $O - E$ indicate that we are only interested in the absolute difference between O and E. This means that we only consider the magnitude of the difference, regardless of whether it is positive or negative.

Formula 49 indicates that for the first cell (brand A) we subtract .5 from the absolute difference, square the remainder, and divide by its expected frequency (E). We then follow the same process for the second cell (brand B), and sum the values for both cells (indicated by Σ) to arrive at the chi-square statistic. The calculation of χ^2, using Formula 49 for the data in Example 18-1, follows.

	O	E	O − E	|O − E| − .5
Brand A	17	25	−8	7.5
Brand B	33	25	8	7.5

Using Formula 49: $\chi^2 = \dfrac{(7.5)^2}{25} + \dfrac{(7.5)^2}{25} = 4.50$

We now need to evaluate this χ^2 using the appropriate chi-square sampling distribution. As was the case with the t distributions and F distributions, there is a family of chi-square distributions, with each distribution based on a specific number of degrees of freedom. However, the degrees of freedom associated with a particular χ^2 test do not depend on the size of the sample, as they did in the t and F distributions, but represent the number of cells in which observed frequencies are "free to vary." For the simplest use of the χ^2 technique, the df are determined by the number of cells minus one. This is because if there are 50 frequencies in the study, any number of them may be assigned to one cell, for example, brand A; that is, the frequencies in this cell are "free to vary." However, once the frequency in that cell is ascertained, the frequency in the other cell is fixed—that is, not free to vary. In the example involving the 50 business executives, any number could have chosen brand A. However, once it is determined that 17 executives actually did so, the number of executives choosing brand B is fixed at 33. Thus, in this example, the frequency in only one cell was free to vary; therefore $df = 1$.

Figure 18-1 shows the shapes of the sampling distributions of chi square for 1, 5, and 15 degrees of freedom. Every possible value of the degrees of freedom has its own distinct curve. Examination of these curves reveals some interesting properties of χ^2 distributions. First, χ^2 values depicted on the horizontal axis are all positive, with $\chi^2 = 0$ as the left-hand limit of the distributions. Second, as the df increases, the shapes of the χ^2 distributions approach the normal curve. Third, if we assume that the null hypothesis is true, we expect the value of chi square to be equal to the df associated with it. That is, for chi square with $df = 5$, we would expect to obtain $\chi^2 = 5$ if the null hypothesis is true. As we have found in other statistical tests, the sampling distribution of χ^2 around this expected value is due to sampling error. The expected chi-square values for $df = 1$, $df = 5$, and $df = 15$ are shown as dotted lines in Figure 18-1.

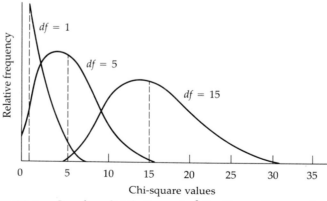

Figure 18-1. Sampling distributions of χ^2 for $df = 1$, $df = 5$, and $df = 15$.

Chi-square tests are always tests of nondirectional hypotheses, in which the right-hand tail of the appropriate χ^2 sampling distribution represents the area for rejection of the null hypothesis. Thus the chi-square technique determines whether there is a significant difference between the O and E frequencies in the cells, not whether a particular cell has a smaller or a larger frequency than expected.

In performing a χ^2 test, we are deciding whether our obtained χ^2 is so large that it is unlikely to have come from the appropriate χ^2 sampling distribution. Large χ^2 values are located in the right-hand tail of the distribution, and they become more improbable as they grow larger. Critical values of χ^2 for the right-hand tail of the various sampling distributions of χ^2 are given in Table 5 at the back of the book. For χ^2 with $df = 1$, a value of 3.84 in needed for significance at $P = .05$. In Example 18-1, we obtained $\chi^2 = 4.50$. Therefore, we reject the null hypothesis that there is no difference in the preferences of business executives for brand A and brand B cigarettes, since the data indicate that brand B is preferred.

Formula 49 shows that the size of the obtained chi square is determined by the size of the discrepancies between O and E in the cells. The evaluation of the significance of a chi square is based on the comparison of the obtained test statistic with the appropriate sampling distribution of χ^2. Since the test statistic is derived from actual sample data, which are discrete measurements, Formula 49 provides for a correction factor that results in a test statistic whose sampling distribution more closely approximates the continuous sampling distribution of chi square. This correction, called *Yates' correction for continuity*, is incorporated in Formula 49 in the form of the subtraction of .5 from the absolute discrepancies between O and E for each cell. Its function is to reduce the discrepancies between O and E and to bring the obtained χ^2, which is based on discrete frequencies, more in line with the continuous function of the sampling distribution. Yates's correction for continuity is appropriate for all chi-square analyses in which $df = 1$.

Where there are more than two categories of frequencies in a chi-square goodness-of-fit test (hence, df is larger than 1), Formula 50 is appropriate for calculating the value of χ^2.

FORMULA 50

Calculation of the chi square when df is larger than 1.

$$\chi^2 = \Sigma \frac{(O - E)^2}{E}$$

df = number of cells $-$ 1

Note that in Formula 50 the Yates' correction for continuity is not applied; it is not necessary when $df > 1$. In such situations, the expected frequency for

EXAMPLE 18-2 Each boy in four random samples of fifth-grade boys was asked to name his favorite superhero. The researcher set $P = .05$ as the level of significance, obtained their responses, and calculated chi square as follows.

	Observed	Expected
Superman	47	35
Batman	38	35
Spiderman	32	35
Aquaman	23	35
	$N = 140$	

Using Formula 50:
$$\chi^2 = \frac{(47-35)^2}{35} + \frac{(38-35)^2}{35} + \frac{(32-35)^2}{35} + \frac{(23-35)^2}{35} = 8.743$$

$$df = 4 - 1 = 3$$

. . . .

Table 5 indicates that a chi square of 7.82 is required for significance at the .05 level. Therefore, the researcher rejects the null hypothesis that there is no difference among the choices of superheroes among fifth-grade boys.

The chi-square test illustrated in Examples 18-1 and 18-2 involves categorizing the data on one variable for a single sample. It is called a goodness-of-fit test because it provides an index of how close the fit is between the observed and expected frequencies, and it is appropriate for testing a variety of hypotheses since it can test the goodness of fit between the observed frequencies and any set of expected frequencies. In Example 18-1, instead of testing the null hypothesis that brand A and brand B were equally preferred by business executives, we could have tested the hypothesis that 75% of them preferred brand A. In this case, we would set 37.5 (75% of 50) as the expected frequency for brand A and 12.5 (25% of 50) as the expected frequency for brand B. We would then use these expected frequencies in Formula 49 to calculate chi square.

THE TEST FOR INDEPENDENCE OF TWO VARIABLES

In the goodness-of-fit test, we employed the chi-square technique to the one-variable situation. Chi square can also be employed to test the hypothesis that the population frequency distribution among the categories on one variable is independent of the distribution on the other variable.

EXAMPLE 18-3   A scout executive wanted to determine if there was a difference between 11-year-old and 14-year-old scouts in their preference for swimming or hiking activities. He randomly selected 28 11-year-old and 30 14-year-old members from the local scout council and did a survey. He set $P = .05$ as his level of significance and obtained the following data:

	Age of scouts		
	11 years	14 years	Total
Hiking	19	12	31
Swimming	9	18	27
Total	28	30	58

· · · · · · ·

Example 18-3 presents data involving the two variables of age and activity preference for a group of scouts. The data given in this example are called contingency data, and the table is referred to as a contingency table. In this example, we want to determine whether the frequency distribution of respondents according to age is independent of the frequency distribution according to activity preference, or whether activity preference is in some way contingent upon the ages of the scouts. The null hypothesis to be tested is that the population distribution of activity preferences is independent of the scouts' ages.

The degrees of freedom in Example 18-3 are $df = 1$ because, given the row and column totals shown, once the frequency in any one of the four cells is ascertained, the frequencies of the other three cells are fixed—that is, they are not free to vary. Thus, if 19 is given as the frequency of 11-year-old scouts choosing hiking, the frequencies in the other three cells must be as they are shown in the example.

In general, the degrees of freedom for a chi-square analysis are determined by: (number of rows $-$ 1) (number of columns $-$ 1). In Example 18-3, where there is a 2 × 2 matrix, $df = (2 - 1)(2 - 1) = 1$. Therefore, Formula 49 is appropriate because it incorporates Yates' correction for continuity.

To apply Formula 49, the first step is to determine the expected frequency for each cell in the matrix. If the null hypothesis that there is no difference in the preferences of 11- and 14-year-old scouts for hiking and swimming activities is true, then, since 31/58 (or 53.4%) of the total group preferred hiking, we would expect that 53.4% of the 11-year-old scouts (which is 53.4% of 28, or 14.9 of them) would choose hiking. For the 14-year-old scouts, we would expect that 53.4% of 30, or 16,034, would choose hiking. These, then, become the expected frequencies for the two categories of hiking.

Because 27/58 (or 46.6%) of the total group preferred swimming, we calculate the expected frequency of 11-year-old scouts preferring this activity to be 46.6% of 28, which is 13.034. The expected frequency of 14-year-old scouts preferring swimming is 46.6% of 30, which is 13.966. As we have seen, expected frequencies in chi-square analyses can be fractional, even though the observed frequencies cannot.

We can summarize the observed and expected frequencies for Example 18-3 as follows:

	Age of scouts				Total	
	11 years		14 years			
	Observed	Expected	Observed	Expected	Observed	Expected
Hiking	19	14.966	12	16.034	31	31
Swimming	9	13.034	18	13.966	27	27
Total	28	28	30	30	58	58

This matrix illustrates the principle that the sum of the expected frequencies must equal the sum of the observed frequencies for each row and column.

To compute χ^2 using the data in this matrix:

Using Formula 49:
$$\chi^2 = \frac{(|19 - 14.966| - .5)^2}{14.966} + \frac{(|9 - 13.034| - .5)^2}{13.034}$$
$$+ \frac{(|12 - 16.034| - .5)^2}{16.034} + \frac{(|18 - 13.966| - .5)^2}{13.966}$$
$$= 3.467$$

In this study, in which $df = 1$ and .05 is the level of significance, Table 5 indicates that the critical value of χ^2 is 3.84. Therefore, we cannot reject the null hypothesis that activity preference is independent of the ages of the scouts. Of course, this finding only applies to the preferences for hiking and swimming among 11- and 14-year-old scouts.

THE TEST FOR EQUALITY OF PROPORTIONS

The use of the chi-square technique is not limited to situations involving only a 2 × 2 contingency table of frequencies. Chi square can also be used to test hypotheses where frequency data are collected on a number of categories of a variable for a number of samples. Example 18-4 presents a study involving multiple samples that have yielded frequencies for multiple categories of a variable.

EXAMPLE 18-4 A researcher wanted to determine if preschool children whose parents were of different socioeconomic levels would have different preference patterns for animals as pets. He randomly selected children whose parents were from three socioeconomic levels and asked each child to indicate his or her favorite pet.

He set $P = .01$ as the level of significance and obtained the following frequency data.

Animal selected	Socioeconomic level			Row total
	High	Middle	Low	
Dog	17	11	7	35
Cat	16	14	9	39
Rabbit	6	11	14	31
Mouse	10	9	16	35
Column Total	49	45	46	140

• • • • • • • •

For this test, the null hypothesis is that the proportion of individuals selecting each category (animals) is the same for each of the populations sampled (socioeconomic levels); that is, it states that the proportion of children selecting dogs will be the same for the high, middle, and low socioeconomic levels. The same hypothesis is made about the proportions of children selecting cats, rabbits, and mice.

Although the null hypothesis deals with the equality of proportions, the determination of chi square does not involve the proportions in its calculation but is based instead on the frequencies given in the contingency table.

Example 18-4 presents a 4 × 3 matrix of frequencies. The degrees of freedom associated with the chi-square test in this example are (rows − 1)(columns − 1) = (4 − 1)(3 − 1) = 6. Since there are more than 1 degrees of freedom, we use Formula 50, the general formula for chi square, which does not employ Yates' correction for continuity. Before using this formula, however, we must compute the expected frequencies for each of the cells. Formula 51 gives a method for calculating the expected frequency for each cell, using row, column, and frequency totals.

FORMULA 51

Calculation of the expected frequency (E) of a cell.

$$E = \frac{(N_{row})(N_{col})}{N_{total}}$$

df = (number of rows − 1)(number of columns − 1)

To illustrate the use of Formula 51, we will compute the expected frequency of the selection of dogs as pets by children with parents in high socioeconomic level. For this cell, $N_{row} = 35$, the total frequency of children in the sample who are high socioeconomic level; $N_{col} = 49$, the total frequency of children selecting dogs as pets; and $N_{total} = 140$.

Using Formula 51: $E = \dfrac{(35)(49)}{140} = 12.25$

Therefore, we would expect that 12.25 children whose parents belong to a high socioeconomic level would select dogs as pets. The expected frequency for each of the other 11 cells in the matrix is computed in the same manner, using Formula 51 and the appropriate row and column totals. These frequencies follow.

Expected frequencies for Example 18-4

Animal selected	Socioeconomic level			Row total
	High	Middle	Low	
Dog	12.25	11.25	11.50	35
Cat	13.65	12.54	12.81	39
Rabbit	10.85	9.96	10.19	31
Mouse	12.25	11.25	11.50	35
Column total	49	45	46	140

We can now employ Formula 50 to compute the chi square, using the observed frequencies in Example 18-4 and the expected frequencies already computed.

Using Formula 50:
$$\chi^2 = \frac{(17 - 12.25)^2}{12.25} + \frac{(11 - 11.25)^2}{11.25} + \frac{(7 - 11.50)^2}{11.50}$$
$$+ \frac{(16 - 13.65)^2}{13.65} + \frac{(14 - 12.54)^2}{12.54} + \frac{(9 - 12.81)^2}{12.81}$$
$$+ \frac{(6 - 10.85)^2}{10.85} + \frac{(11 - 9.96)^2}{9.96} + \frac{(14 - 10.19)^2}{10.19}$$
$$+ \frac{(10 - 12.25)^2}{12.25} + \frac{(9 - 11.25)^2}{11.25} + \frac{(16 - 11.50)^2}{11.50}$$
$$= 11.65$$

Table 5 indicates that for $df = 6$, a chi square of 16.81 is needed to be significant at the .01 level. Therefore, the obtained $\chi^2 = 11.65$ is not large enough to permit the researcher to reject the null hypothesis in Example 18-4. Thus the researcher concludes that the data does not permit him to say that different socioeconomic levels have different preference patterns.

The chi-square technique is a very useful statistical tool, because it can be used with any number of samples divided into any number of categories of responses. Chi-square tests are not limited solely to nominal data. In Example 18-4, the variable of socioeconomic class was divided into three categories: low, middle, and high. This variable is essentially ordinal in nature (although it can even be measured on the interval scale, if the measurement instrument is calibrated into equal intervals), but for chi-square purposes, it was considered to consist of three categories. The chi-square technique requires only that the frequency of responses be assigned to specific categories of a variable, regardless of whether the nominal, the ordinal, the interval, or the ratio level of measurement is used.

Even though chi-square is a highly versatile nonparametric technique that makes no assumptions about population values, certain requirements must be met in using it properly to analyze data. These requirements are:

1. The sample or samples must have been randomly selected. This requirement applies to all statistical techniques.

2. Each response must be independent of the other responses in the study; that is, the way in which one response is categorized must in no way influence the way in which the other responses are categorized. In Example 18-4, we assumed that the choice of a pet by one child had no effect on the choice of a pet by any other child. Implicit in this requirement of independence is the assumption that each frequency must represent a different individual.

3. Each cell must have an expected frequency of at least 5 when the df equals 1. When the df is greater than 1, this requirement may be relaxed somewhat without invalidating the chi-square procedure. Some authorities suggest that when the df is 2 or more, at least 80% of the cells should have expected frequencies of 5 or more. Note that the preceding discussion relates to expected frequencies, not to observed frequencies.

The reason for these requirements is that, if the observed frequencies are obtained from a multitude of samples where the null hypothesis is true, these frequencies tend to be normally distributed around the expected frequency. This is not the case, however, when the expected frequency is very small. When small expected frequencies occur, the approximation to the chi-square distribution is not very good.

EXERCISES

1. A statistics instructor wanted to determine if students differed in their preference for two types of instructional techniques. He asked a random sample of 96 students whether they preferred programmed instruction or a conventional text. He obtained the following responses. Setting $P = .01$ as the level of significance, should the null hypothesis be rejected?

Type of instruction	f
Programmed instruction	57
Conventional text	39

2. A toy manufacturer wanted to know if there was a difference between boys and girls in their preference for two types of swimming pool floats. He selected a random sample of children and asked them to choose between

type A and type B floats. The following data were obtained. Setting $P = .05$ as the level of significance, should the null hypothesis be rejected?

	Type A	Type B
Boys	30	12
Girls	15	19

3. A social psychologist asked random samples of adults in three socioeconomic levels to specify their preferences among four types of television programs. She obtained the following responses. If $P = .01$ is set as the level of significance, should the null hypothesis be rejected?

Socioeconomic level	Drama	Documentary	Comedy	Musical
High	14	17	9	8
Middle	8	18	10	12
Low	10	9	17	9

4. A researcher wanted to determine if students in urban high schools differed from students in rural high schools in their preference for night football games. He selected two random samples and obtained the following responses. Setting $P = .05$ as the level of significance, should the null hypothesis be rejected?

High school	Prefer night football	
	Yes	No
Urban	6	13
Rural	21	12

5. An experimental psychologist wanted to determine if rats had a preference for eating out of red, yellow, or blue bowls. He randomly selected a sample of rats and observed the following choices. Setting $P = .05$ as the level of significance, should the null hypothesis be rejected?

Bowl color	f
Red	13
Yellow	47
Blue	22

6. A preschool director wanted to determine if working and nonworking mothers differed in their preference for the scheduling of preschool classes for their children. She obtained the following responses. Setting $P = .01$ as the level of significance, should the null hypothesis be rejected?

Parents	Time of classes		
	Morning	Early afternoon	Late afternoon
Working	10	36	14
Nonworking	23	17	7

19 NONPARAMETRIC TECHNIQUES—ORDINAL DATA

This chapter presents some selected nonparametric techniques that are analogs to some of the parametric techniques presented in earlier chapters. The techniques are appropriate whenever the assumption of normality underlying their parametric counterparts cannot be met, or where the data are in the form of ranks. As we will illustrate, even when the data are on the interval or ratio scales, to use the nonparametric techniques we first must convert them to ordinal measurement in the form of ranks. This process of ranking of interval data causes us to lose some of the finer distinctions among the scores. For instance, scores of 92, 73, and 71 would receive ranks of 1, 2, and 3, without regard to the fact that 19 points separate the first two scores, whereas there are only two points between the second and the third scores. This loss of information may lead a nonparametric technique to yield a nonsignificant finding, whereas a parametric technique, if appropriate, might result in a significant finding.

Three major factors should be considered when addressing the question of whether to use a parametric or nonparametric technique for analyzing statistical data. (There is no choice, of course, where the data are on the nominal or ordinal scale. Here nonparametrics must be used.)

1. When the assumption of normality holds, parametrics are generally the preferred type because they are generally more powerful than nonparametrics. That is, they more often lead us to reject a false null hypothesis when it is, in fact, false. This reduces the probability of making a Type II error.

2. Even when the assumptions cannot be completely met, parametric tests have been shown to be quite *robust*. That is, they tend to provide accurate results even if distributions depart substantially from being normally distributed, particularly if large samples are used.

3. Parametric and nonparametric techniques produce somewhat different types of findings. Whereas a parametric test may tell whether one mean score is different from another mean score, the nonparametric counterpart examines whether one distribution is in any way different from the other distribution. Thus these two types of techniques are somewhat sensitive to different aspects of the distributions; this may lead one type to a significant finding and the other type to one of nonsignificance, when applied to the same data.

In the past, the complexity of performing hand calculations using parametric techniques was a factor, nonparametrics being so much easier to use. However, with the advent of computer analysis procedures, this is no longer of concern.

Many nonparametric techniques are available to the statistician. In addition to chi square, presented in Chapter 18, we present here five useful ones for ordinal data that are analogs to parametric techniques as follows:

Parametric test	Nonparametric test
Pearson correlation (r)	Spearman rank correlation (rho)
t test—Independent samples	Mann-Whitney U test
t test—Dependent samples	Wilcoxon matched-pairs signed-ranks test
One-way analysis of variance	Kruskal-Wallis H test
Two-way analysis of variance	Friedman two-way ANOVA for ranks

SPEARMAN RANK CORRELATION

This correlation gives us a measure of the relationship between two variables where the data are in the form of relative rankings of individuals on each variable. In such cases, the data represent measurements on the ordinal scale. The Spearman rank correlation is represented by the Greek letter rho (ρ).

Example 19-1 gives a typical correlational study in which rho is the appropriate technique for data analysis.

EXAMPLE 19-1 A high school counselor wanted to determine if there is a relationship (either positive or negative) between students' athletic ability and their level of anxiety. He ranked 12 students according to their level of anxiety as he perceived them during counseling sessions. He asked the physical education instructor to rank these students according to their athletic ability. He set $P = .05$ as the level of significance. The following data were obtained:

Student	Athletic ability rank	Anxiety level rank	Student	Athletic ability rank	Anxiety level rank
A	1	4	G	7	10
B	2	3	H	8	8
C	3	1	I	10	7
D	4.5	6	J	10	9
E	4.5	5	K	10	12
F	6	2	L	12	11

∙ ∙ ∙ ∙ ∙ ∙ ∙ ∙

The data given in Example 19-1 on the two variables, athletic ability and anxiety, represent the rank order of the students. It is customary, though not necessary, to assign the rank of 1 to the individual with the highest score on a variable. Where more than one individual occupies the same position in the order—that is, where two or more are "tied"—each is assigned the average of the ranks that he would otherwise have received.

In Example 19-1, student A is rated highest in athletic ability and is assigned the rank of 1. Students D and E are rated as having the same degree of athletic ability. Since they occupy the fourth and fifth positions in the rank order, each of them is assigned the average of these two ranks, which is 4.5.

The rankings in Example 19-1 indicate that there is a tendency for students who are ranked high on one variable to be ranked high on the other variable also. To make a statistical test of this relationship, we proceed in a manner similar to that used in calculating the Pearson product-moment correlation coefficient. The null hypothesis to be tested can be stated as: "There is no relationship between the variables of athletic ability and anxiety in high school students." Our task is to answer the statistical question: "What is the probability that the obtained relationship between the rankings of a sample of individuals on these two variables is a result of sampling error?" In Example 19-1, the counselor set .05 as the level of significance. In this study, the counselor is interested in detecting a relationship, either positive or negative, between the two variables. Therefore, a nondirectional hypothesis test is appropriate.

The statistical technique appropriate for testing the null hypothesis in this example is the Spearman rank correlation. The null hypothesis is rho = 0. Formula 52 shows the method of computation for this correlation coefficient.

FORMULA 52

Calculation of Spearman's rank correlation coefficient (rho).

$$\text{rho} = 1 - \frac{6 \Sigma D^2}{N(N^2 - 1)}$$

where: D = difference between a pair of ranks
N = number of pairs of ranks

Formula 52 indicates that we must determine ΣD^2, which is obtained by taking the difference between each student's two rankings, squaring each difference, and then summing the squares. In Formula 52, the value 6 is a constant, and N is the number of pairs of rankings, or the number of individuals in the study. The computation of rho for the data in Example 19-1 is shown in Table 19-1.

Table 6 in the back of the book is used to evaluate the significance of rho, when N is 30 or less. A procedure for determining significance when N is 30

TABLE 19-1. Computation for rho for data in Example 19-1

Student	Athletic ability ranking	Anxiety level ranking	D	D²
A	1	4	3	9
B	2	3	1	1
C	3	1	−2	4
D	4.5	6	1.5	2.25
E	4.5	5	.5	.25
F	6	2	−4	16
G	7	10	3	9
H	8	8	0	0
I	10	7	−3	9
J	10	9	−1	1
K	10	12	2	4
L	12	11	−1	1
N = 12				$\Sigma D^2 = 56.5$

Using Formula 52: $\text{rho} = 1 - \dfrac{6(56.5)}{12(12^2 - 1)} = .801$

or more is given later in this chapter. Note that the first column on the left-hand side is designated N, the number of pairs of ranks in the correlation. The concept of degrees of freedom does not apply when we use this statistical technique.

The values of rho given in the body of Table 6 represent the critical values at varying significance levels for both directional and nondirectional hypothesis tests.

In Example 19-1, where the counselor designated .05 as the significance level for making a nondirectional hypothesis test, and where $N = 12$, Table 6 indicates that a rho of .588 or larger is needed to reject the null hypothesis. The obtained rho = .801 is therefore large enough to permit the rejection of the null hypothesis. Thus the data indicate that students with exceptional athletic ability tend to be more anxious than students with low athletic ability. The counselor may then conclude that there is a positive relationship between athletic ability and anxiety levels of high school students.

The Spearman rank correlation is also commonly used when the assumption that the variables are normally distributed in the population is unwarranted. Such an instance is described in Example 19-2.

EXAMPLE 19-2 A special education teacher gave each of the eight educable mentally retarded students in the class a manual dexterity test. The teacher hypothesized that there is a positive relationship between the students' manual dexterity and their

IQ scores. The level of significance was set at .01. The teacher obtained the following data.

Student	Manual dexterity scores	IQ scores
A	21	84
B	6	70
C	18	84
D	22	85
E	9	84
F	4	73
G	21	79
H	20	75

....

The data in Example 19-2 reveal that manual dexterity scores cannot be assumed to be normally distributed in the population of educable mentally retarded students. Also we cannot assume that IQ scores are normally distributed in this particular population. When testing a hypothesis, if either or both variables cannot meet the assumption of normality, the parametric Pearson product-moment correlation is not warranted, and we must turn to the nonparametric rank-order correlation.

To apply this technique, we must first convert the scores on each variable to rankings. For the variable of manual dexterity, we assign a rank of 1 to student F, who has the lowest score. The other manual dexterity scores are ranked in order of magnitude. The same procedure is followed for the IQ scores. Students A, C, and E all have the same IQ and occupy ranks 5, 6, and 7. Accordingly, each is assigned the average rank of 6. The rankings of the scores on each variable and the computation of rho appear in Table 19-2.

TABLE 19-2. Computation of rho for data in Example 19-2

Student	Manual dexterity		IQ		D	D^2
	Score	Rank	Score	Rank		
A	21	6.5	84	6	− .5	.25
B	6	2	70	1	−1	1
C	18	4	84	6	2	4
D	22	8	85	8	0	0
E	9	3	84	6	3	9
F	4	1	73	2	1	1
G	21	6.5	79	4	−2.5	6.25
H	20	5	75	3	−2	4
N = 8						$\Sigma D^2 = 25.5$

Using Formula 52: $\text{rho} = 1 - \dfrac{6(25.5)}{8(8^2 - 1)} = .696$

Because the teacher hypothesized that there would be a positive correlation between these two variables in the population, this is a directional hypothesis, and so a one-tail test is made. Table 6 indicates that the critical value of rho at $P = .01$ for a one-tail test is .953. Our obtained correlation rho = .696 is not large enough to permit us to reject the null hypothesis, and we must conclude that this correlation may be due solely to sampling error.

Of course, the validity of our decision regarding the null hypothesis rests on the assumption that the individuals in the sample have been randomly selected and are representative (within sampling error) of the population we want to generalize to. This is true of all statistical tests. In Examples 19-1 and 19-2, both samples were deliberately kept small for computational convenience as have all samples in this text. In reality, much larger samples would be needed to provide definitive findings. (It is highly unlikely that the eight students in the educable mentally retarded class are representative of all such students!)

Table 6 gives the critical values for directional and nondirectional tests for sample sizes of 30 or less. Where N exceeds 30 the t distribution can be used to evaluate the significance of the rank correlation coefficient. To use this distribution, we obtain the t ratio from Formula 53, with degrees of freedom determined by the number of pairs of ranks minus 2.

FORMULA 53

Computation of t to evaluate significance of rho when $N > 30$.

$$t = \frac{\text{rho}\sqrt{N-2}}{\sqrt{1-\text{rho}^2}}$$

$$df = N - 2$$

where: N = number of pairs of ranks.

To illustrate the use of Formula 52, suppose that a researcher obtains a rho of .764 between two variables, using a sample of 50 subjects. For these data,

$$t = \frac{.764\sqrt{50-2}}{\sqrt{1-(.764)^2}}$$

$$t = \frac{5.293}{.645} = 8.206$$

degrees of freedom = $50 - 2 = 48$

Table 2 indicates that for $df = 60$ (closest tabled value to 48), a t ratio of 2.000 is required for significance at $p = .05$ (two-tailed test). Since the obtained t ratio is larger than the critical value, the researcher should conclude that there is a significant correlation between the two variables.

THE MANN-WHITNEY U TEST

The nonparametric technique named the Mann-Whitney U Test is a test of the null hypothesis that there is no difference in the distribution of scores of the populations from which two samples are selected. This test is the nonparametric alternative to the t test for independent samples and is relatively easy to perform. It only assumes that the two samples are independent and that the data are theoretically from continuous variables.

To use the Mann-Whitney U Test we must calculate the value of U and evaluate it for significance. To do this, we must first rank all of the scores in both samples combined, assigning a rank of 1 to the smallest score in the combined samples. Then the ranks assigned to scores in each sample are summed, with $R1$ being assigned to the sum of ranks for sample 1 and $R2$ to the sum of ranks for sample 2. The values $U1$ and $U2$ are then calculated using Formula 54.

FORMULA 54

Calculation of $U1$ and $U2$ for the Mann-Whitney U Test.

$$U1 = N_1 N_2 + \frac{N_1(N_1 + 1)}{2} - R1$$

$$U2 = N_1 N_2 + \frac{N_2(N_2 + 1)}{2} - R2$$

where: N_1 = size of sample 1
N_2 = size of sample 2
$R1$ = sum of ranks assigned to sample 1
$R2$ = sum of ranks assigned to sample 2

The test of significance is applied to U, which is defined as the smaller of $U1$ and $U2$.

Example 19-3 is presented to illustrate the use of the Mann-Whitney U test.

EXAMPLE 19-3 A psychologist wished to determine if there was a difference between ninth-grade boys and girls in their attitudes toward school. She developed an inventory to measure this variable and obtained attitude scores for ten boys and eight girls. She set $P = .05$ as the level of significance.

Not having any knowledge about how scores on this inventory would be distributed in the population, she selected the nonparametric technique, the Mann-Whitney U test, to analyze the data. The data and her calculations are as follows.

Boys		Girls	
Attitude scores	Rank	Attitude scores	Rank
22	3	29	5
30	6	32	7
20	1.5	37	10
36	9	39	11
20	1.5	35	8
40	12	41	13
27	4	44	18
42	14	43	16
43	16	$R_2 = 88$	
43	16	$N_2 = 8$	
$R1 = 83$			
$N_1 = 10$			

Using Formula 54: $U1 = 10(8) + \dfrac{10(10+1)}{2} - 83 = 52$

$U2 = 10(8) + \dfrac{8(8-1)}{2} - 88 = 28$

· · · · · · · ·

Note that the attitude scores are ranked across both groups, and the ranks assigned to boys are summed as $R1$ and the ranks assigned to girls are summed as $R2$. As usual, all scores with the same value receive the average of the ranks they would have occupied had the ties not occurred. Also, note that the sample with the higher sum of ranks receives the smaller U. The $U2$ of 28, being the smaller of the two, is taken for U in testing the null hypothesis.

Table 7 in the appendix presents the critical values for U at six commonly used significance levels. This table may be used whenever the sample sizes of both groups are 20 or less. The table is entered by locating the larger sample size across the top of the table (column headings) and the smaller sample size along the left margin (row headings). The values in the body of the table are the critical values of U. These critical values of U are for directional hypotheses (one-tail tests). (When testing a nondirectional hypothesis, the probability values need to be doubled.)

If the U obtained using Formula 54 is *smaller* than the tabled critical value, the null hypothesis is rejected. It is important to emphasize that the evaluation of U is just the opposite of the procedure used up to now, where an obtained value had to *exceed* the critical value in order to be significant.

In Example 19-3, when $N_1 = 10$ and $N_2 = 8$, Table 7 gives the critical value of U as 21 at $P = .05$. Since the psychologist wants to test a nondirectional hypothesis, we need to double the P values in Table 7. For $N = 10$ and $N = 8$, Table 7 gives the critical value of U as 24 (listed for $P = .025$, which must be doubled for $P = .05$ in a two-tail test). The obtained value of $U = 28$ is larger than the critical value, so the psychologist cannot reject the null hypothesis of no difference between the attitude scores of boys and girls.

When the size of the samples exceeds 20, the sampling distribution of U approaches the normal distribution. To evaluate the significance of U for large samples, we need to calculate μ_u and σ_u, using Formulas 55 and 56, and then compute the z ratio using Formula 57.

FORMULA 55

Calculation of the population mean of U.

$$\mu_u = \frac{N_1 N_2}{2}$$

FORMULA 56

Calculation of the population standard deviation of U.

$$\sigma_u = \sqrt{\frac{(N_1)(N_2)(N_1 + N_2 + 1)}{12}}$$

FORMULA 57

Calculation of the z ratio in the Mann-Whitney U test.

$$z = \frac{U - \mu_u}{\sigma_u}$$

We will not illustrate the straightforward calculations required by Formulas 55 through 57. The obtained z ratio is evaluated for statistical significance using the probabilities associated with the normal curve as given in Table 1.

THE WILCOXON MATCHED-PAIRS SIGNED-RANKS TEST

The nonparametric technique that is used when there are matched pairs of data is the Wilcoxon matched-pairs signed-ranks test. It tests the hypothesis that there is no difference in the distributions of the matched populations from which the sample data are derived. The test is commonly applied to data in a pretest-posttest research design when the assumptions underlying its parametric analog, the *t* test for dependent samples, cannot be met. The calculation procedure is simple:

1. Obtain the difference between the pair of scores for each subject. (If a difference score is zero, drop the pair of data from the analysis.)
2. Rank the absolute values (without regard to their algebraic sign) of the difference scores, assigning the rank of 1 to the smallest absolute difference.
3. Attach the algebraic sign of each difference score to its rank. (This is why this test is called "signed-ranks.")
4. Sum the ranks of the positive ranks.
5. Sum the ranks of the negative ranks.

The smaller of the two sums of ranks is designated as T. This is the value that is compared with the critical value of T. Table 8 in the appendix provides the critical values of T for directional and nondirectional tests when the sample size is 50 or less. Note that if the obtained T is equal to or *less* than the critical value, the null hypothesis is rejected. Whenever the sample size exceeds 50, the distribution of T approximates the normal distribution.

Formulas 58 and 59 give methods for calculating μ_T and σ_T. Formula 60 provides the z ratio. This z ratio is evaluated in the normal manner, using Table 1. If the obtained z ratio exceeds the critical value, the null hypothesis is rejected.

FORMULA 58

Calculation of the population mean of T.

$$\mu_T = \frac{N(N+1)}{4}$$

FORMULA 59

Calculation of the population standard deviation of T.

$$\sigma_T = \sqrt{\frac{N(N+1)(2N+1)}{24}}$$

FORMULA 60

Calculation of the z ratio in the Wilcoxon test.

$$z = \frac{T - \mu_T}{\sigma_T}$$

The use of the Wilcoxon matched-pairs signed-ranks test is presented in Example 19-4.

EXAMPLE 19-4 An instructor of a computer course wanted to determine if there was a significant increase in the amount of knowledge acquired about computer operations as a result of a two-week class. He gave a pretest prior to instruction and a posttest at the conclusion of the class and obtained the following test scores. He set $P = .05$ as the level of significance.

Pretest score	Posttest score	Difference	Absolute difference	Rank of absolute difference	Signed ranks
59	72	13	13	7	7
61	59	−2	2	2	−2
70	74	4	4	3.5	3.5
32	57	25	25	8	8
40	41	1	1	1	1
39	35	−4	4	3.5	−3.5
72	77	5	5	5	5
51	58	7	7	6	6

$T\text{(positive)} = 30.5$
$T\text{(negative)} = 5.5$

• • • • • • •

The table of data in Example 19-4 shows all the steps necessary to calculate the two T values. First, the difference between the scores in each pair is obtained. Then the absolute differences are ranked. Next the appropriate sign is attached to each rank. Finally the positive and negative ranks are summed separately. The smaller T is 5.5, so it is the T to be evaluated.

Table 8 indicates that for $N = 8$ at $P = .05$ for a two-tail test, the critical value of T is 3. In Example 19-4, the instructor's obtained T of 5.5 exceeds this critical value, so the null hypothesis is not rejected. He cannot conclude that there was a significant increase in scores from pretest to posttest. Of course, he had a very small sample in which almost everyone would have to show an increase for T to be significant. For example, if everyone had gained in score over time, there would be no negative ranks. Hence T would be zero, which would have been significant.

THE KRUSKAL-WALLIS ONE-WAY ANALYSIS OF VARIANCE BY RANKS TEST

This technique is the nonparametric analog to the one-way analysis of variance for independent samples. It is useful whenever the usual assumptions underlying the analysis of variance cannot be met. The hypothesis that it tests is that there is no difference in the distribution of scores of the various populations from which the samples were selected.

The procedure for performing this test is similar to that of other tests in this chapter. The scores in the various samples combined are ranked, with a rank of 1 being assigned to the smallest score. The sum of the ranks, designated R_j, is obtained for each sample, and H is calculated by using Formula 61.

FORMULA 61

Calculation of H in the Kruskal-Wallis test.

$$H = \frac{12}{N(N+1)} \Sigma \frac{R_j^2}{n_j} - 3(N+1)$$

where: R_j is the sum of ranks in a given sample
n_j is the size of a given sample
N is the total number in all samples

The sampling distribution of H follows the chi-square distribution with degrees of freedom being the number of samples minus one ($k-1$).

The use of the Kruskal-Wallis technique is illustrated in Example 19-5.

EXAMPLE 19-5 A book publisher wanted to determine if length of sales experience differentially affected the volume of sales of the salespeople on her staff. She set $P = .01$ as the level of significance. She grouped her sales force into three categories of experience and obtained the following sales data.

Little experience		Some experience		Much experience	
Sales	Rank	Sales	Rank	Sales	Rank
12	3	19	7.5	34	15
17	6	22	9	29	13
15	4	16	5	36	16
19	7.5	28	11.5	39	18
9	2	27	10	37	17
7	1	28	11.5	$R_3 = 79$	
$R_1 = 23.5$		30	14		
		$R_2 = 68.5$			

Using Formula 61: $H = \dfrac{12}{18(18+1)} \left[\dfrac{(23.5)^2}{6} + \dfrac{(68.5)^2}{7} + \dfrac{(79)^2}{5} \right] - 3(18+1) = 13.57$

• • • • • • • •

The degrees of freedom for Example 19-5 is $3 - 1 = 2$. Using Table 5, the publisher finds that at $df = 2$, a chi square of 9.21 is required for significance at $P = .01$. The obtained chi square of 13.57 clearly exceeds the critical value, so the publisher rejects the null hypothesis and concludes that length of experience does indeed affect the quantity of sales.

THE FRIEDMAN TWO-WAY ANALYSIS OF VARIANCE BY RANKS

This technique is available as a nonparametric test based on ranks that is applicable when subjects have been matched or when repeated measurements are taken on the same subjects or groups of subjects. It provides a test of the null hypothesis that the related samples of measures came from populations with identical distributions. As such, it is an analog to the parametric one-way analysis of variance for repeated measures. It is a useful technique whenever repeated measures are taken. These measures can represent either individual responses or a group's response, such as a mean, median, or sum of responses.

To use the Friedman test, the data are arranged in a two-way table having N rows and k columns. Each row contains the scores of each subject (or group) for the various conditions represented by the columns. The conditions are the categories of the independent variable under study. Thus each row gives the scores of one subject for the k conditions.

The data are converted to ordinal measurement by ranking each subject's scores across the conditions. The ranks assigned to each of the conditions are then summed and are used in Formula 62.

FORMULA 62

Calculation of chi square in the Friedman two-way ANOVA test.

$$\chi^2 = \frac{12}{Nk(k+1)} \Sigma (R_j)^2 - 3N(k+1)$$

where: N is the number of subjects
k is the number of conditions
R_j is the sum of ranks for a given condition

Example 19-6 illustrates the ranking process and the calculation of chi square using the Friedman two-way ANOVA technique.

EXAMPLE 19-6

A marketing agent had the hypothesis that customers perceive different colored cereal boxes as containing different quantities of cereal, when, in fact, they all contain equal quantities. She ran an experiment in which eight subjects were asked to estimate the quantity, in ounces, in each of three different colored boxes randomly presented. She set $P = .01$ as the level of significance and obtained the following data.

	Conditions		
	Red	Yellow	Blue
Subject 1	15	12	9
Subject 2	16	17	15
Subject 3	10	10	8
Subject 4	14	15	17
Subject 5	13	12	10
Subject 6	12	11	9
Subject 7	13	10	10
Subject 8	12	12	11

Ranking each subject's responses across the three conditions, she obtained the following ranks:

	Red	Yellow	Blue
Subject 1	3	2	1
Subject 2	2	3	1
Subject 3	2.5	2.5	1
Subject 4	1	2	3
Subject 5	3	2	1
Subject 6	3	2	1
Subject 7	3	1.5	1.5
Subject 8	2.5	2.5	1
	$R_1 = 20$	$R_2 = 17.5$	$R_3 = 10.5$

....

In this example, $N = 8$ and $k = 3$. Note that for each subject the conditions are ranked from 1 (lowest) to 3 (highest) value. Tied values are handled in the usual manner. (In this technique, it does not matter whether 1 is assigned to the smallest or largest value.) The sum of ranks for each column is designated as R.

If the null hypothesis is correct, the sum of the ranks should be approximately equal. To the extent that these sums differ, the chi-square value is increased, which may result in a significant finding.

Applying Formula 62 to the preceding data, we obtain:

$$\chi^2 = \frac{12}{8(3)(3+1)}[(20)^2 + (17.5)^2 + (10.5)^2] - 3(8)(3+1) = 6.06$$

To evaluate the significance of the obtained chi square, we use one of two tables, depending on the size of the samples and the number of conditions. For $k = 3$ with N between 2 and 9, and for $k = 4$ with N between 2 and 4, Table 9 provides the probabilities associated with various chi-square values. This table should be consulted whenever appropriate. In Example 19-6 the obtained chi square of 6.06, with $N = 8$ and $k = 3$, has a P between .079 and .047 (tabled probabilities for chi-square values of 5.25 and 4.75, respectively). Since the obtained chi square clearly is not large enough to be significant at the preset .01 level, the null hypothesis should not be rejected. This illustrates how difficult it is to obtain a significant finding when using small samples.

Where large samples are used, the obtained chi square is evaluated for significance, in the usual manner, by consulting Table 5, with $k - 1$ degrees of freedom.

EXERCISES

For the following exercises, no assumptions can be made regarding the shape of the population distributions.

1. Four samples of college freshmen were given a writing proficiency test after receiving different methods of writing instruction. The level of significance was set at $P = .01$. Should the null hypothesis that there is no difference in the effectiveness of the four methods of writing instruction be rejected?

Method A	Method B	Method C	Method D
14	15	9	7
15	16	11	10
17	16	12	11
17	16	14	12
19	17	12	14
	14	12	15
	12		

2. Two samples of seventh-grade students were matched on the basis of their IQs; one group was given form A of a social studies test and the other group was given form B of the same test. The test developers wanted to test the hypothesis that there was a positive relationship between the two sets of data. They set $P = .05$ as the level of significance. Should the null hypothesis be rejected?

Form A	Form B	Form A	Form B
141	139	145	143
154	146	149	150
149	157	145	139
160	157	132	130
145	145	129	134
156	159		

3. A developmental psychologist wanted to test the hypothesis that there is a difference in the scores that pairs of different-sex twins receive on a spatial relations test. She selected 14 pairs of twins and obtained the following test data. Setting $P = .01$ as the level of significance, should the null hypothesis be rejected?

Male twin	Female twin	Male twin	Female twin
25	26	23	19
18	17	27	32
20	24	24	23
17	19	24	21
25	31	32	27
21	19	24	29
33	34	29	29

4. A researcher had the hypothesis that there would be a difference among the mother, father, and teacher ratings of children's self-concept. Using a self-concept scale, she obtained the following data. She set $P = .05$ as the level of significance. Should the null hypothesis be rejected?

Mother	Father	Teacher
23	17	22
21	18	23
20	20	21
26	21	24
28	22	27
24	24	25
25	26	27
29	27	32
36	29	37
38	29	37

5. A signal corps instructor made the hypothesis that spaced instruction was more effective than concentrated instruction for increasing Morse code proficiency in recruits. He randomly assigned recruits to the two methods of instruction and administered a Morse code proficiency test to them. He set $P = .01$ as the level of significance and obtained the following data. Should he reject the null hypothesis?

Spaced instruction	Concentrated instruction
24	31
20	37
19	32
23	33
17	29
19	38
18	39
30	37
32	
27	

6. A high school athletic coach wanted to test the hypothesis that there was no relationship between swimming ability and height in students. She ranked the students on their swimming ability, assigning a rank of 1 to the best swimmer, and measured their heights. She set $P = .05$ as the level of significance. From the following data, should the null hypothesis be rejected?

Swimming ranks	Heights in inches	Swimming ranks	Heights in inches
1	68	4	70
3	66	5.5	70
5.5	63	8	65
2	70	9	69
7	64		

7. A camp director randomly selected three samples of Girl Scouts and gave each sample a different style of leadership during a summer encampment. At the conclusion of the camping period, each Girl Scout answered a questionnaire designed to measure her attitude toward the camp. The camp director obtained the following attitude measurements. She set $P = .05$ as the level of significance. Should the null hypothesis be rejected?

Democratic	Authoritarian	Laissez-faire
17	18	17
16	15	14
16	15	13
15	11	12
11		10
		8

8. A biologist randomly assigned white rats to two groups to examine the hypothesis: "There will be a difference between the life spans of white rats living in a type M environment and the life spans of white rats living in a type N environment." She selected $P = .05$ as the level of significance and obtained the following data. Should the null hypothesis be rejected?

Type M environment	Type N environment
12	8
20	10
18	19
16	17
17	12
15	16
19	16
13	15
13	14

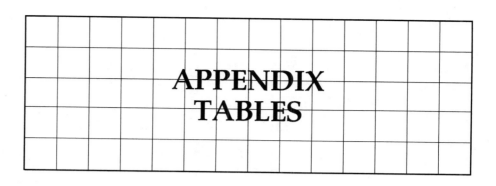

1. Areas of the Standard Normal Curve
2. Critical Values of t
3. Critical Values of F
4. Critical Values of Q
5. Critical Values of Chi Square
6. Critical Values for the Spearman Rank-Order Correlation Coefficient
7. Critical Value of U for the Mann–Whitney U Test
8. Critical Values of W for the Wilcoxon Test
9. Table of Probabilities

TABLE 1. AREAS OF THE STANDARD NORMAL CURVE

z	μ to z	z	μ to z	z	μ to z	z	μ to z	z	μ to z	z	μ to z
0.00	.0000	0.25	.0987	0.50	.1915	0.75	.2734	1.00	.3413	1.25	.3944
0.01	.0040	0.26	.1026	0.51	.1950	0.76	.2764	1.01	.3438	1.26	.3962
0.02	.0080	0.27	.1064	0.52	.1985	0.77	.2794	1.02	.3461	1.27	.3980
0.03	.0120	0.28	.1103	0.53	.2019	0.78	.2823	1.03	.3485	1.28	.3997
0.04	.0160	0.29	.1141	0.54	.2054	0.79	.2852	1.04	.3508	1.29	.4015
0.05	.0199	0.30	.1179	0.55	.2088	0.80	.2881	1.05	.3531	1.30	.4032
0.06	.0239	0.31	.1217	0.56	.2123	0.81	.2910	1.06	.3554	1.31	.4049
0.07	.0279	0.32	.1255	0.57	.2157	0.82	.2939	1.07	.3577	1.32	.4066
0.08	.0319	0.33	.1293	0.58	.2190	0.83	.2967	1.08	.3599	1.33	.4082
0.09	.0359	0.34	.1331	0.59	.2221	0.84	.2995	1.09	.3621	1.34	.4099
0.10	.0398	0.35	.1368	0.60	.2257	0.85	.3023	1.10	.3643	1.35	.4115
0.11	.0438	0.36	.1406	0.61	.2291	0.86	.3051	1.11	.3665	1.36	.4131
0.12	.0478	0.37	.1443	0.62	.2324	0.87	.3078	1.12	.3686	1.37	.4147
0.13	.0517	0.38	.1480	0.63	.2357	0.88	.3106	1.13	.3708	1.38	.4162
0.14	.0557	0.39	.1517	0.64	.2389	0.89	.3133	1.14	.3729	1.39	.4177
0.15	.0596	0.40	.1554	0.65	.2422	0.90	.3159	1.15	.3749	1.40	.4192
0.16	.0636	0.41	.1591	0.66	.2454	0.91	.3186	1.16	.3770	1.41	.4207
0.17	.0675	0.42	.1628	0.67	.2486	0.92	.3212	1.17	.3790	1.42	.4222
0.18	.0714	0.43	.1664	0.68	.2517	0.93	.3238	1.18	.3810	1.43	.4236
0.19	.0753	0.44	.1700	0.69	.2549	0.94	.3264	1.19	.3830	1.44	.4251
0.20	.0793	0.45	.1736	0.70	.2580	0.95	.3289	1.20	.3849	1.45	.4265
0.21	.0832	0.46	.1772	0.71	.2611	0.96	.3315	1.21	.3869	1.46	.4279
0.22	.0871	0.47	.1808	0.72	.2642	0.97	.3340	1.22	.3888	1.47	.4292
0.23	.0910	0.48	.1844	0.73	.2673	0.98	.3365	1.23	.3907	1.48	.4306
0.24	.0948	0.49	.1879	0.74	.2704	0.99	.3389	1.24	.3925	1.49	.4319
0.25	.0987	0.50	.1915	0.75	.2734	1.00	.3413	1.25	.3944	1.50	.4332

z	μ to z	z	μ to z
1.50	.4332	1.75	.4599
1.51	.4345	1.76	.4608
1.52	.4357	1.77	.4616
1.53	.4370	1.78	.4625
1.54	.4382	1.79	.4633
1.55	.4394	1.80	.4641
1.56	.4406		
1.57	.4418		
1.58	.4429		
1.59	.4441		
1.60	.4452		
1.61	.4463		
1.62	.4474		
1.63	.4484		
1.64	.4495		
1.65	.4505		
1.66	.4515		
1.67	.4525		
1.68	.4535		
1.69	.4545		
1.70	.4554		
1.71	.4564		
1.72	.4573		
1.73	.4582		
1.74	.4591		
1.75	.4599		

Adapted from Table 1, I. E. S. Pearson and H. O. Hartley (Eds.), *Biometrika Tables for Statisticians* (3rd Ed.). Copyright 1966. Reprinted by permission of the Biometrika Trustees.

z	μ to z	z	μ to z	z	μ to z	z	μ to z	z	μ to z
1.80	.4641	2.05	.4798	2.30	.4893	2.55	.4946	2.80	.4974	3.05	.4989
1.81	.4649	2.06	.4803	2.31	.4896	2.56	.4948	2.81	.4975	3.06	.4989
1.82	.4656	2.07	.4808	2.32	.4898	2.57	.4949	2.82	.4976	3.07	.4989
1.83	.4664	2.08	.4812	2.33	.4901	2.58	.4951	2.83	.4977	3.08	.4990
1.84	.4671	2.09	.4817	2.34	.4904	2.59	.4952	2.84	.4977	3.09	.4990
1.85	.4678	2.10	.4821	2.35	.4906	2.60	.4953	2.85	.4978	3.10	.4990
1.86	.4686	2.11	.4826	2.36	.4909	2.61	.4955	2.86	.4979	3.11	.4991
1.87	.4693	2.12	.4830	2.37	.4911	2.62	.4956	2.87	.4979	3.12	.4991
1.88	.4699	2.13	.4834	2.38	.4913	2.63	.4957	2.88	.4980	3.13	.4991
1.89	.4706	2.14	.4838	2.39	.4916	2.64	.4959	2.89	.4981	3.14	.4992
1.90	.4713	2.15	.4842	2.40	.4918	2.65	.4960	2.90	.4981	3.15	.4992
1.91	.4719	2.16	.4846	2.41	.4920	2.66	.4961	2.91	.4982	3.16	.4992
1.92	.4726	2.17	.4850	2.42	.4922	2.67	.4962	2.92	.4982	3.17	.4992
1.93	.4732	2.18	.4854	2.43	.4925	2.68	.4963	2.93	.4983	3.18	.4993
1.94	.4738	2.19	.4857	2.44	.4927	2.69	.4964	2.94	.4984	3.19	.4993
1.95	.4744	2.20	.4861	2.45	.4929	2.70	.4965	2.95	.4984	3.20	.4993
1.96	.4750	2.21	.4864	2.46	.4931	2.71	.4966	2.96	.4985	3.21	.4993
1.97	.4756	2.22	.4868	2.47	.4932	2.72	.4967	2.97	.4985	3.22	.4994
1.98	.4761	2.23	.4871	2.48	.4934	2.73	.4968	2.98	.4986	3.23	.4994
1.99	.4767	2.24	.4875	2.49	.4936	2.74	.4969	2.99	.4986	3.24	.4994
2.00	.4772	2.25	.4878	2.50	.4938	2.75	.4970	3.00	.4987	3.30	.4995
2.01	.4778	2.26	.4881	2.51	.4940	2.76	.4971	3.01	.4987	3.40	.4997
2.02	.4783	2.27	.4884	2.52	.4941	2.77	.4972	3.02	.4987	3.50	.4998
2.03	.4788	2.28	.4887	2.53	.4943	2.78	.4973	3.03	.4988	3.60	.4998
2.04	.4793	2.29	.4890	2.54	.4945	2.79	.4974	3.04	.4988	3.70	.4999
2.05	.4798	2.30	.4893	2.55	.4946	2.80	.4974	3.05	.4989		

TABLE 2. CRITICAL VALUES OF t

df	α Levels for Two-Tailed Test					
	.2	.1	.05	.02	.01	.001
1	3.078	6.314	12.706	31.821	63.657	636.619
2	1.886	2.920	4.303	6.965	9.925	31.598
3	1.638	2.353	3.182	4.541	5.841	12.924
4	1.533	2.132	2.776	3.747	4.604	8.610
5	1.476	2.015	2.571	3.365	4.032	6.869
6	1.440	1.943	2.447	3.143	3.707	5.959
7	1.415	1.895	2.365	2.998	3.499	5.408
8	1.397	1.860	2.306	2.896	3.355	5.041
9	1.383	1.833	2.262	2.821	3.250	4.781
10	1.372	1.812	2.228	2.764	3.169	4.587
11	1.363	1.796	2.201	2.718	3.106	4.437
12	1.356	1.782	2.179	2.681	3.055	4.318
13	1.350	1.771	2.160	2.650	3.012	4.221
14	1.345	1.761	2.145	2.624	2.977	4.140
15	1.341	1.753	2.131	2.602	2.947	4.073
16	1.337	1.746	2.120	2.583	2.921	4.015
17	1.333	1.740	2.110	2.567	2.898	3.965
18	1.330	1.734	2.101	2.552	2.878	3.922
19	1.328	1.729	2.093	2.539	2.861	3.883
20	1.325	1.725	2.086	2.528	2.845	3.850
21	1.323	1.721	2.080	2.518	2.831	3.819
22	1.321	1.717	2.074	2.508	2.819	3.792
23	1.319	1.714	2.069	2.500	2.807	3.767
24	1.318	1.711	2.064	2.492	2.797	3.745
25	1.316	1.708	2.060	2.485	2.787	3.725
26	1.315	1.706	2.056	2.479	2.779	3.707
27	1.314	1.703	2.052	2.473	2.771	3.690
28	1.313	1.701	2.048	2.467	2.763	3.674
29	1.311	1.699	2.045	2.462	2.756	3.659
30	1.310	1.697	2.042	2.457	2.750	3.646
40	1.303	1.684	2.021	2.423	2.704	3.551
60	1.296	1.671	2.000	2.390	2.660	3.460
120	1.289	1.658	1.980	2.358	2.617	3.373
∞	1.282	1.645	1.960	2.326	2.576	3.291
	.10	.05	.025	.01	.005	.0005
	α Levels for a One-Tailed Test					

Source: This table is taken from Table III of Fisher and Yates, *Statistical Tables for Biological, Agricultural and Medical Research*, published by Longman Group Ltd., London (previously published by Oliver and Boyd, Ltd., Edinburgh), and by permission of the authors and publishers.
Note: Reject null hypothesis if obtained t ratio is equal to or greater than tabled value.

TABLE 3. CRITICAL VALUES OF F

.05 Level (roman) and .01 Level (bold face) α Levels for the Distribution of F

Degrees of Freedom (for the numerator)

	1	2	3	4	5	6	7	8	9	10	11	12	14	16	20	24	30	40	50	75	100	200	500	∞
1	161 **4,052**	200 **4,999**	216 **5,403**	225 **5,625**	230 **5,764**	234 **5,859**	237 **5,928**	239 **5,981**	241 **6,022**	242 **6,056**	243 **6,082**	244 **6,106**	245 **6,142**	246 **6,169**	248 **6,208**	249 **6,234**	250 **6,258**	251 **6,286**	252 **6,302**	253 **6,323**	253 **6,334**	254 **6,352**	254 **6,361**	254 **6,366**
2	18.51 **98.49**	19.00 **99.00**	19.16 **99.17**	19.25 **99.25**	19.30 **99.30**	19.33 **99.33**	19.36 **99.34**	19.37 **99.36**	19.38 **99.38**	19.39 **99.40**	19.40 **99.41**	19.41 **99.42**	19.42 **99.43**	19.43 **99.44**	19.44 **99.45**	19.45 **99.46**	19.46 **99.47**	19.47 **99.48**	19.47 **99.48**	19.48 **99.49**	19.49 **99.49**	19.49 **99.49**	19.50 **99.50**	19.50 **99.50**
3	10.13 **34.12**	9.55 **30.82**	9.28 **29.46**	9.12 **28.71**	9.01 **28.24**	8.94 **27.91**	8.88 **27.67**	8.84 **27.49**	8.81 **27.34**	8.78 **27.23**	8.76 **27.13**	8.74 **27.05**	8.71 **26.92**	8.69 **26.83**	8.66 **26.69**	8.64 **26.60**	8.62 **26.50**	8.60 **26.41**	8.58 **26.35**	8.57 **26.27**	8.56 **26.23**	8.54 **26.18**	8.54 **26.14**	8.53 **26.12**
4	7.71 **21.20**	6.94 **18.00**	6.59 **16.69**	6.39 **15.98**	6.26 **15.52**	6.16 **15.21**	6.09 **14.98**	6.04 **14.80**	6.00 **14.66**	5.96 **14.54**	5.93 **14.45**	5.91 **14.37**	5.87 **14.24**	5.84 **14.15**	5.80 **14.02**	5.77 **13.93**	5.74 **13.83**	5.71 **13.74**	5.70 **13.69**	5.68 **13.61**	5.66 **13.57**	5.65 **13.52**	5.64 **13.48**	5.63 **13.46**
5	6.61 **16.26**	5.79 **13.27**	5.41 **12.06**	5.19 **11.39**	5.05 **10.97**	4.95 **10.67**	4.88 **10.45**	4.82 **10.27**	4.78 **10.15**	4.74 **10.05**	4.70 **9.96**	4.68 **9.89**	4.64 **9.77**	4.60 **9.68**	4.56 **9.55**	4.53 **9.47**	4.50 **9.38**	4.46 **9.29**	4.44 **9.24**	4.42 **9.17**	4.40 **9.13**	4.38 **9.07**	4.37 **9.04**	4.36 **9.02**
6	5.99 **13.74**	5.14 **10.92**	4.76 **9.78**	4.53 **9.15**	4.39 **8.75**	4.28 **8.47**	4.21 **8.26**	4.15 **8.10**	4.10 **7.98**	4.06 **7.87**	4.03 **7.79**	4.00 **7.72**	3.96 **7.60**	3.92 **7.52**	3.87 **7.39**	3.84 **7.31**	3.81 **7.23**	3.77 **7.14**	3.75 **7.09**	3.72 **7.02**	3.71 **6.99**	3.69 **6.94**	3.68 **6.90**	3.67 **6.88**
7	5.59 **12.25**	4.74 **9.55**	4.35 **8.45**	4.12 **7.85**	3.97 **7.46**	3.87 **7.19**	3.79 **7.00**	3.73 **6.84**	3.68 **6.71**	3.63 **6.62**	3.60 **6.54**	3.57 **6.47**	3.52 **6.35**	3.49 **6.27**	3.44 **6.15**	3.41 **6.07**	3.38 **5.98**	3.34 **5.90**	3.32 **5.85**	3.29 **5.78**	3.28 **5.75**	3.25 **5.70**	3.24 **5.67**	3.23 **5.65**
8	5.32 **11.26**	4.46 **8.65**	4.07 **7.59**	3.84 **7.01**	3.69 **6.63**	3.58 **6.37**	3.50 **6.19**	3.44 **6.03**	3.39 **5.91**	3.34 **5.82**	3.31 **5.74**	3.28 **5.67**	3.23 **5.56**	3.20 **5.48**	3.15 **5.36**	3.12 **5.28**	3.08 **5.20**	3.05 **5.11**	3.03 **5.06**	3.00 **5.00**	2.98 **4.96**	2.96 **4.91**	2.94 **4.88**	2.93 **4.86**
9	5.12 **10.56**	4.26 **8.02**	3.86 **6.99**	3.63 **6.42**	3.48 **6.06**	3.37 **5.80**	3.29 **5.62**	3.23 **5.47**	3.18 **5.35**	3.13 **5.26**	3.10 **5.18**	3.07 **5.11**	3.02 **5.00**	2.98 **4.92**	2.93 **4.80**	2.90 **4.73**	2.86 **4.64**	2.82 **4.56**	2.80 **4.51**	2.77 **4.45**	2.76 **4.41**	2.73 **4.36**	2.72 **4.33**	2.71 **4.31**
10	4.96 **10.04**	4.10 **7.56**	3.71 **6.55**	3.48 **5.99**	3.33 **5.64**	3.22 **5.39**	3.14 **5.21**	3.07 **5.06**	3.02 **4.95**	2.97 **4.85**	2.94 **4.78**	2.91 **4.71**	2.86 **4.60**	2.82 **4.52**	2.77 **4.41**	2.74 **4.33**	2.70 **4.25**	2.67 **4.17**	2.64 **4.12**	2.61 **4.05**	2.59 **4.01**	2.56 **3.96**	2.55 **3.93**	2.54 **3.91**
11	4.84 **9.65**	3.98 **7.20**	3.59 **6.22**	3.36 **5.67**	3.20 **5.32**	3.09 **5.07**	3.01 **4.88**	2.95 **4.74**	2.90 **4.63**	2.86 **4.54**	2.82 **4.46**	2.79 **4.40**	2.74 **4.29**	2.70 **4.21**	2.65 **4.10**	2.61 **4.02**	2.57 **3.94**	2.53 **3.86**	2.50 **3.80**	2.47 **3.74**	2.45 **3.70**	2.42 **3.66**	2.41 **3.62**	2.40 **3.60**
12	4.75 **9.33**	3.88 **6.93**	3.49 **5.95**	3.26 **5.41**	3.11 **5.06**	3.00 **4.82**	2.92 **4.65**	2.85 **4.50**	2.80 **4.39**	2.76 **4.30**	2.72 **4.22**	2.69 **4.16**	2.64 **4.05**	2.60 **3.98**	2.54 **3.86**	2.50 **3.78**	2.46 **3.70**	2.42 **3.61**	2.40 **3.56**	2.36 **3.49**	2.35 **3.46**	2.32 **3.41**	2.31 **3.38**	2.30 **3.36**
13	4.67 **9.07**	3.80 **6.70**	3.41 **5.74**	3.18 **5.20**	3.02 **4.86**	2.92 **4.62**	2.84 **4.44**	2.77 **4.30**	2.72 **4.19**	2.67 **4.10**	2.63 **4.02**	2.60 **3.96**	2.55 **3.85**	2.51 **3.78**	2.46 **3.67**	2.42 **3.59**	2.38 **3.51**	2.34 **3.42**	2.32 **3.37**	2.28 **3.30**	2.26 **3.27**	2.24 **3.21**	2.22 **3.18**	2.21 **3.16**

Degrees of Freedom (for the denominator)

Source: Reproduced by permission from *Statistical Methods*, 5th edition by George B. Snedecor, copyright 1956 by the Iowa State University Press.
Note: Reject null hypothesis if F ratio is equal to or greater than tabled value.

206 APPENDIX TABLES

TABLE 3. (CONTINUED)

Degrees of Freedom (for the numerator)

	1	2	3	4	5	6	7	8	9	10	11	12	14	16	20	24	30	40	50	75	100	200	500	∞	
14	4.60 8.86	3.74 6.51	3.34 5.56	3.11 5.03	2.96 4.69	2.85 4.46	2.77 4.28	2.70 4.14	2.65 4.03	2.60 3.94	2.56 3.86	2.53 3.80	2.48 3.70	2.44 3.62	2.39 3.51	2.35 3.43	2.31 3.34	2.27 3.26	2.24 3.21	2.21 3.14	2.19 3.11	2.16 3.06	2.14 3.02	2.13 3.00	14
15	4.54 8.68	3.68 6.36	3.29 5.42	3.06 4.89	2.90 4.56	2.79 4.32	2.70 4.14	2.64 4.00	2.59 3.89	2.55 3.80	2.51 3.73	2.48 3.67	2.43 3.56	2.39 3.48	2.33 3.36	2.29 3.29	2.25 3.20	2.21 3.12	2.18 3.07	2.15 3.00	2.12 2.97	2.10 2.92	2.08 2.89	2.07 2.87	15
16	4.49 8.53	3.63 6.23	3.24 5.29	3.01 4.77	2.85 4.44	2.74 4.20	2.66 4.03	2.59 3.89	2.54 3.78	2.49 3.69	2.45 3.61	2.42 3.55	2.37 3.45	2.33 3.37	2.28 3.25	2.24 3.18	2.20 3.10	2.16 3.01	2.13 2.96	2.09 2.89	2.07 2.86	2.04 2.80	2.02 2.77	2.01 2.75	16
17	4.45 8.40	3.59 6.11	3.20 5.18	2.96 4.67	2.81 4.34	2.70 4.10	2.62 3.93	2.55 3.79	2.50 3.68	2.45 3.59	2.41 3.52	2.38 3.45	2.33 3.35	2.29 3.27	2.23 3.16	2.19 3.08	2.15 3.00	2.11 2.92	2.08 2.86	2.04 2.79	2.02 2.76	1.99 2.70	1.97 2.67	1.96 2.65	17
18	4.41 8.28	3.55 6.01	3.16 5.09	2.93 4.58	2.77 4.25	2.66 4.01	2.58 3.85	2.51 3.71	2.46 3.60	2.41 3.51	2.37 3.44	2.34 3.37	2.29 3.27	2.25 3.19	2.19 3.07	2.15 3.00	2.11 2.91	2.07 2.83	2.04 2.78	2.00 2.71	1.89 2.68	1.95 2.62	1.93 2.59	1.92 2.57	18
19	4.38 8.18	3.52 5.93	3.13 5.01	2.90 4.50	2.74 4.17	2.63 3.94	2.55 3.77	2.48 3.63	2.43 3.52	2.38 3.43	2.34 3.36	2.31 3.30	2.26 3.19	2.21 3.12	2.15 3.00	2.11 2.92	2.07 2.84	2.02 2.76	2.00 2.70	1.96 2.63	1.94 2.60	1.91 2.54	1.90 2.51	1.88 2.49	19
20	4.35 8.10	3.49 5.85	3.10 4.94	2.87 4.43	2.71 4.10	2.60 3.87	2.52 3.71	2.45 3.56	2.40 3.45	2.35 3.37	2.31 3.30	2.28 3.23	2.23 3.13	2.18 3.05	2.12 2.94	2.08 2.86	2.04 2.77	1.99 2.69	1.96 2.63	1.92 2.56	1.90 2.53	1.87 2.47	1.85 2.44	1.84 2.42	20
21	4.32 8.02	3.47 5.78	3.07 4.87	2.84 4.37	2.68 4.04	2.57 3.81	2.49 3.65	2.42 3.51	2.37 3.40	2.32 3.31	2.28 3.24	2.25 3.17	2.20 3.07	2.15 2.99	2.09 2.88	2.05 2.80	2.00 2.72	1.96 2.63	1.93 2.58	1.89 2.51	1.87 2.47	1.84 2.42	1.82 2.38	1.81 2.36	21
22	4.30 7.94	3.44 5.72	3.05 4.82	2.82 4.31	2.66 3.99	2.55 3.76	2.47 3.59	2.40 3.45	2.35 3.35	2.30 3.26	2.26 3.18	2.23 3.12	2.18 3.02	2.13 2.94	2.07 2.83	2.03 2.75	1.98 2.67	1.93 2.58	1.91 2.53	1.87 2.46	1.84 2.42	1.81 2.37	1.80 2.33	1.78 2.31	22
23	4.28 7.88	3.42 5.66	3.03 4.76	2.80 4.26	2.64 3.94	2.53 3.71	2.45 3.54	2.38 3.41	2.32 3.30	2.28 3.21	2.24 3.14	2.20 3.07	2.14 2.97	2.10 2.89	2.04 2.78	2.00 2.70	1.96 2.62	1.91 2.53	1.88 2.48	1.84 2.41	1.82 2.37	1.79 2.32	1.77 2.28	1.76 2.26	23
24	4.26 7.82	3.40 5.61	3.01 4.72	2.78 4.22	2.62 3.90	2.51 3.67	2.43 3.50	2.36 3.36	2.30 3.25	2.26 3.17	2.22 3.09	2.18 3.03	2.13 2.93	2.09 2.85	2.02 2.74	1.98 2.66	1.94 2.58	1.89 2.49	1.86 2.44	1.82 2.36	1.80 2.33	1.76 2.27	1.74 2.23	1.73 2.21	24
25	4.24 7.77	3.38 5.57	2.99 4.68	2.76 4.18	2.60 3.86	2.49 3.63	2.41 3.46	2.34 3.32	2.28 3.21	2.24 3.13	2.20 3.05	2.16 2.99	2.11 2.89	2.06 2.81	2.00 2.70	1.96 2.62	1.92 2.54	1.87 2.45	1.84 2.40	1.80 2.32	1.77 2.29	1.74 2.23	1.72 2.19	1.71 2.17	25
26	4.22 7.72	3.37 5.53	2.98 4.64	2.74 4.14	2.59 3.82	2.47 3.59	2.39 3.42	2.32 3.29	2.27 3.17	2.22 3.09	2.18 3.02	2.15 2.96	2.10 2.86	2.05 2.77	1.99 2.66	1.95 2.58	1.90 2.50	1.85 2.41	1.82 2.36	1.78 2.28	1.76 2.25	1.72 2.19	1.70 2.15	1.69 2.13	26

Degrees of Freedom (for the denominator)

TABLE 3. (CONTINUED)

	\multicolumn{17}{c}{Degrees of Freedom (for the numerator)}	
	1	2	3	4	5	6	7	8	9	10	11	12	14	16	20	24	30	40	50	75	100	200	500	∞	
27	4.21 7.68	3.35 5.49	2.96 4.60	2.73 4.11	2.57 3.79	2.46 3.56	2.37 3.39	2.30 3.26	2.25 3.14	2.20 3.06	2.16 2.98	2.13 2.93	2.08 2.83	2.03 2.74	1.97 2.63	1.93 2.55	1.88 2.47	1.84 2.38	1.80 2.33	1.76 2.25	1.74 2.21	1.71 2.16	1.68 2.12	1.67 2.10	27
28	4.20 7.64	3.34 5.45	2.95 4.57	2.71 4.07	2.56 3.76	2.44 3.53	2.36 3.36	2.29 3.23	2.24 3.11	2.19 3.03	2.15 2.95	2.12 2.90	2.06 2.80	2.02 2.71	1.96 2.60	1.91 2.52	1.87 2.44	1.81 2.35	1.78 2.30	1.75 2.22	1.72 2.18	1.69 2.13	1.67 2.09	1.65 2.06	28
29	4.18 7.60	3.33 5.42	2.93 4.54	2.70 4.04	2.54 3.73	2.43 3.50	2.35 3.33	2.28 3.20	2.22 3.08	2.18 3.00	2.14 2.92	2.10 2.87	2.05 2.77	2.00 2.68	1.94 2.57	1.90 2.49	1.85 2.41	1.80 2.32	1.77 2.27	1.73 2.19	1.71 2.15	1.68 2.10	1.65 2.06	1.64 2.03	29
30	4.17 7.56	3.32 5.39	2.92 4.51	2.69 4.02	2.53 3.70	2.42 3.47	2.34 3.30	2.27 3.17	2.21 3.06	2.16 2.98	2.12 2.90	2.09 2.84	2.04 2.74	1.99 2.66	1.93 2.55	1.89 2.47	1.84 2.38	1.79 2.29	1.76 2.24	1.72 2.16	1.69 2.13	1.66 2.07	1.64 2.03	1.62 2.01	30
32	4.15 7.50	3.30 5.34	2.90 4.46	2.67 3.97	2.51 3.66	2.40 3.42	2.32 3.25	2.25 3.12	2.19 3.01	2.14 2.94	2.10 2.86	2.07 2.80	2.02 2.70	1.97 2.62	1.91 2.51	1.86 2.42	1.82 2.34	1.76 2.25	1.74 2.20	1.69 2.12	1.67 2.08	1.64 2.02	1.61 1.98	1.59 1.96	32
34	4.13 7.44	3.28 5.29	2.88 4.42	2.65 3.93	2.49 3.61	2.38 3.38	2.30 3.21	2.23 3.08	2.17 2.97	2.12 2.89	2.08 2.82	2.05 2.76	2.00 2.66	1.95 2.58	1.89 2.47	1.84 2.38	1.80 2.30	1.74 2.21	1.71 2.15	1.67 2.08	1.64 2.04	1.61 1.98	1.59 1.94	1.57 1.91	34
36	4.11 7.39	3.26 5.25	2.86 4.38	2.63 3.89	2.48 3.58	2.36 3.35	2.28 3.18	2.21 3.04	2.15 2.94	2.10 2.86	2.06 2.78	2.03 2.72	1.98 2.62	1.93 2.54	1.87 2.43	1.82 2.35	1.78 2.26	1.72 2.17	1.69 2.12	1.65 2.04	1.62 2.00	1.59 1.94	1.56 1.90	1.55 1.87	36
38	4.10 7.35	3.25 5.21	2.85 4.34	2.62 3.86	2.46 3.54	2.35 3.32	2.26 3.15	2.19 3.02	2.14 2.91	2.09 2.82	2.05 2.75	2.02 2.69	1.96 2.59	1.92 2.51	1.85 2.40	1.80 2.32	1.76 2.22	1.71 2.14	1.67 2.08	1.63 2.00	1.60 1.97	1.57 1.90	1.54 1.86	1.53 1.84	38
40	4.08 7.31	3.23 5.18	2.84 4.31	2.61 3.83	2.45 3.51	2.34 3.29	2.25 3.12	2.18 2.99	2.12 2.88	2.07 2.80	2.04 2.73	2.00 2.66	1.95 2.56	1.90 2.49	1.84 2.37	1.79 2.29	1.74 2.20	1.69 2.11	1.66 2.05	1.61 1.97	1.59 1.94	1.55 1.88	1.53 1.84	1.51 1.81	40
42	4.07 7.27	3.22 5.15	2.83 4.29	2.59 3.80	2.44 3.49	2.32 3.26	2.24 3.10	2.17 2.96	2.11 2.86	2.06 2.77	2.02 2.70	1.99 2.64	1.94 2.54	1.89 2.46	1.82 2.35	1.78 2.26	1.73 2.17	1.68 2.08	1.64 2.02	1.60 1.94	1.57 1.91	1.54 1.85	1.51 1.80	1.49 1.78	42
44	4.06 7.24	3.21 5.12	2.82 4.26	2.58 3.78	2.43 3.46	2.31 3.24	2.23 3.07	2.16 2.94	2.10 2.84	2.05 2.75	2.01 2.68	1.98 2.62	1.92 2.52	1.88 2.44	1.81 2.32	1.76 2.24	1.72 2.15	1.66 2.06	1.63 2.00	1.58 1.92	1.56 1.88	1.52 1.82	1.50 1.78	1.48 1.75	44
46	4.05 7.21	3.20 5.10	2.81 4.24	2.57 3.76	2.42 3.44	2.30 3.22	2.22 3.05	2.14 2.92	2.09 2.82	2.04 2.73	2.00 2.66	1.97 2.60	1.91 2.50	1.87 2.42	1.80 2.30	1.75 2.22	1.71 2.13	1.65 2.04	1.62 1.98	1.57 1.90	1.54 1.86	1.51 1.80	1.48 1.76	1.46 1.72	46
48	4.04 7.19	3.19 5.08	2.80 4.22	2.56 3.74	2.41 3.42	2.30 3.20	2.21 3.04	2.14 2.90	2.08 2.80	2.03 2.71	1.99 2.64	1.96 2.58	1.90 2.48	1.86 2.40	1.79 2.28	1.74 2.20	1.70 2.11	1.64 2.02	1.61 1.96	1.56 1.88	1.53 1.84	1.50 1.78	1.47 1.73	1.45 1.70	48

Degrees of Freedom (for the denominator)

TABLE 3. (CONTINUED)

Degrees of Freedom (for the numerator)

	1	2	3	4	5	6	7	8	9	10	11	12	14	16	20	24	30	40	50	75	100	200	500	∞	
50	4.03 7.17	3.18 5.06	2.79 4.20	2.56 3.72	2.40 3.41	2.29 3.18	2.20 3.02	2.13 2.88	2.07 2.78	2.02 2.70	1.98 2.62	1.95 2.56	1.90 2.46	1.85 2.39	1.78 2.26	1.74 2.18	1.69 2.10	1.63 2.00	1.60 1.94	1.55 1.86	1.52 1.82	1.48 1.76	1.46 1.71	1.44 1.68	50
55	4.02 7.12	3.17 5.01	2.78 4.16	2.54 3.68	2.38 3.37	2.27 3.15	2.18 2.98	2.11 2.85	2.05 2.75	2.00 2.66	1.97 2.59	1.93 2.53	1.88 2.43	1.83 2.35	1.76 2.23	1.72 2.15	1.67 2.06	1.61 1.96	1.58 1.90	1.52 1.82	1.50 1.78	1.46 1.71	1.43 1.66	1.41 1.64	55
60	4.00 7.08	3.15 4.98	2.76 4.13	2.52 3.65	2.37 3.34	2.25 3.12	2.17 2.95	2.10 2.82	2.04 2.72	1.99 2.63	1.95 2.56	1.92 2.50	1.86 2.40	1.81 2.32	1.75 2.20	1.70 2.12	1.65 2.03	1.59 1.93	1.56 1.87	1.50 1.79	1.48 1.74	1.44 1.68	1.41 1.63	1.39 1.60	60
65	3.99 7.04	3.14 4.95	2.75 4.10	2.51 3.62	2.36 3.31	2.24 3.09	2.15 2.93	2.08 2.79	2.02 2.70	1.98 2.61	1.94 2.54	1.90 2.47	1.85 2.37	1.80 2.30	1.73 2.18	1.68 2.09	1.63 2.00	1.57 1.90	1.54 1.84	1.49 1.76	1.46 1.71	1.42 1.64	1.39 1.60	1.37 1.56	65
70	3.98 7.01	3.13 4.92	2.74 4.08	2.50 3.60	2.35 3.29	2.23 3.07	2.14 2.91	2.07 2.77	2.01 2.67	1.97 2.59	1.93 2.51	1.89 2.45	1.84 2.35	1.79 2.28	1.72 2.15	1.67 2.07	1.62 1.98	1.56 1.88	1.53 1.82	1.47 1.74	1.45 1.69	1.40 1.62	1.37 1.56	1.35 1.53	70
80	3.96 6.96	3.11 4.88	2.72 4.04	2.48 3.56	2.33 3.25	2.21 3.04	2.12 2.87	2.05 2.74	1.99 2.64	1.95 2.55	1.91 2.48	1.88 2.41	1.82 2.32	1.77 2.24	1.70 2.11	1.65 2.03	1.60 1.94	1.54 1.84	1.51 1.78	1.45 1.70	1.42 1.65	1.38 1.57	1.35 1.52	1.32 1.49	80
100	3.94 6.90	3.09 4.82	2.70 3.98	2.46 3.51	2.30 3.20	2.19 2.99	2.10 2.82	2.03 2.69	1.97 2.59	1.92 2.51	1.88 2.43	1.85 2.36	1.79 2.26	1.75 2.19	1.68 2.06	1.63 1.98	1.57 1.89	1.51 1.79	1.48 1.73	1.42 1.64	1.39 1.59	1.34 1.51	1.30 1.46	1.28 1.43	100
125	3.92 6.84	3.07 4.78	2.68 3.94	2.44 3.47	2.29 3.17	2.17 2.95	2.08 2.79	2.01 2.65	1.95 2.56	1.90 2.47	1.86 2.40	1.83 2.33	1.77 2.23	1.72 2.15	1.65 2.03	1.60 1.94	1.55 1.85	1.49 1.75	1.45 1.68	1.39 1.59	1.36 1.54	1.31 1.46	1.27 1.40	1.25 1.37	125
150	3.91 6.81	3.06 4.75	2.67 3.91	2.43 3.44	2.27 3.14	2.16 2.92	2.07 2.76	2.00 2.62	1.94 2.53	1.89 2.44	1.85 2.37	1.82 2.30	1.76 2.20	1.71 2.12	1.64 2.00	1.59 1.91	1.54 1.83	1.47 1.72	1.44 1.66	1.37 1.56	1.34 1.51	1.29 1.43	1.25 1.37	1.22 1.33	150
200	3.89 6.76	3.04 4.71	2.65 3.88	2.41 3.41	2.26 3.11	2.14 2.90	2.05 2.73	1.98 2.60	1.92 2.50	1.87 2.41	1.83 2.34	1.80 2.28	1.74 2.17	1.69 2.09	1.62 1.97	1.57 1.88	1.52 1.79	1.45 1.69	1.42 1.62	1.35 1.53	1.32 1.48	1.26 1.39	1.22 1.33	1.19 1.28	200
400	3.86 6.70	3.02 4.66	2.62 3.83	2.39 3.36	2.23 3.06	2.12 2.85	2.03 2.69	1.96 2.55	1.90 2.46	1.85 2.37	1.81 2.29	1.78 2.23	1.72 2.12	1.67 2.04	1.60 1.92	1.54 1.84	1.49 1.74	1.42 1.64	1.38 1.57	1.32 1.47	1.28 1.42	1.22 1.32	1.16 1.24	1.13 1.19	400
1000	3.85 6.66	3.00 4.62	2.61 3.80	2.38 3.34	2.22 3.04	2.10 2.82	2.02 2.66	1.95 2.53	1.89 2.43	1.84 2.34	1.80 2.26	1.76 2.20	1.70 2.09	1.65 2.01	1.58 1.89	1.53 1.81	1.47 1.71	1.41 1.61	1.36 1.54	1.30 1.44	1.26 1.38	1.19 1.28	1.13 1.19	1.08 1.11	1000
∞	3.84 6.64	2.99 4.60	2.60 3.78	2.37 3.32	2.21 3.02	2.09 2.80	2.01 2.64	1.94 2.51	1.88 2.41	1.83 2.32	1.79 2.24	1.75 2.18	1.69 2.07	1.64 1.99	1.57 1.87	1.52 1.79	1.46 1.69	1.40 1.59	1.35 1.52	1.28 1.41	1.24 1.36	1.17 1.25	1.11 1.15	1.00 1.00	∞

Degrees of Freedom (for the denominator)

TABLE 4. CRITICAL VALUES OF Q

d.f.	P	m=2	3	4	5	6	7	8	9	10	11
5	.05	3.64	4.60	5.22	5.67	6.03	6.33	6.58	6.80	6.99	7.17
	.01	5.70	6.98	7.80	8.42	8.91	9.32	9.67	9.97	10.24	10.48
6	.05	3.46	4.34	4.90	5.30	5.63	5.90	6.12	6.32	6.49	6.65
	.01	5.24	6.33	7.03	7.56	7.97	8.32	8.61	8.87	9.10	9.30
7	.05	3.34	4.16	4.68	5.06	5.36	5.61	5.82	6.00	6.16	6.30
	.01	4.95	5.92	6.54	7.01	7.37	7.68	7.94	8.17	8.37	8.55
8	.05	3.26	4.04	4.53	4.89	5.17	5.40	5.60	5.77	5.92	6.05
	.01	4.75	5.64	6.20	6.62	6.96	7.24	7.47	7.68	7.86	8.03
9	.05	3.20	3.95	4.41	4.76	5.02	5.24	5.43	5.59	5.74	5.87
	.01	4.60	5.43	5.96	6.35	6.66	6.91	7.13	7.33	7.49	7.65
10	.05	3.15	3.88	4.33	4.65	4.91	5.12	5.30	5.46	5.60	5.72
	.01	4.48	5.27	5.77	6.14	6.43	6.67	6.87	7.05	7.21	7.36
11	.05	3.11	3.82	4.26	4.57	4.82	5.03	5.20	5.35	5.49	5.61
	.01	4.39	5.15	5.62	5.97	6.25	6.48	6.67	6.84	6.99	7.13
12	.05	3.08	3.77	4.20	4.51	4.75	4.95	5.12	5.27	5.39	5.51
	.01	4.32	5.05	5.50	5.84	6.10	6.32	6.51	6.67	6.81	6.94
13	.05	3.06	3.73	4.15	4.45	4.69	4.88	5.05	5.19	5.32	5.43
	.01	4.26	4.96	5.40	5.73	5.98	6.19	6.37	6.53	6.67	6.79
14	.05	3.03	3.70	4.11	4.41	4.64	4.83	4.99	5.13	5.25	5.36
	.01	4.21	4.89	5.32	5.63	5.88	6.08	6.26	6.41	6.54	6.66
15	.05	3.01	3.67	4.08	4.37	4.59	4.78	4.94	5.08	5.20	5.31
	.01	4.17	4.84	5.25	5.56	5.80	5.99	6.16	6.31	6.44	6.55
16	.05	3.00	3.65	4.05	4.33	4.56	4.74	4.90	5.03	5.15	5.26
	.01	4.13	4.79	5.19	5.49	5.72	5.92	6.08	6.22	6.35	6.46
17	.05	2.98	3.63	4.02	4.30	4.52	4.70	4.86	4.99	5.11	5.21
	.01	4.10	4.74	5.14	5.43	5.66	5.85	6.01	6.15	6.27	6.38
18	.05	2.97	3.61	4.00	4.28	4.49	4.67	4.82	4.96	5.07	5.17
	.01	4.07	4.70	5.09	5.38	5.60	5.79	5.94	6.08	6.20	6.31
19	.05	2.96	3.59	3.98	4.25	4.47	4.65	4.79	4.92	5.04	5.14
	.01	4.05	4.67	5.05	5.33	5.55	5.73	5.89	6.02	6.14	6.25
20	.05	2.95	3.58	3.96	4.23	4.45	4.62	4.77	4.90	5.01	5.11
	.01	4.02	4.64	5.02	5.29	5.51	5.69	5.84	5.97	6.09	6.19
24	.05	2.92	3.53	3.90	4.17	4.37	4.54	4.68	4.81	4.92	5.01
	.01	3.96	4.55	4.91	5.17	5.37	5.54	5.69	5.81	5.92	6.02
30	.05	2.89	3.49	3.85	4.10	4.30	4.46	4.60	4.72	4.82	4.92
	.01	3.89	4.45	4.80	5.05	5.24	5.40	5.54	5.65	5.76	5.85
40	.05	2.86	3.44	3.79	4.04	4.23	4.39	4.52	4.63	4.73	4.82
	.01	3.82	4.37	4.70	4.93	5.11	5.26	5.39	5.50	5.60	5.69
60	.05	2.83	3.40	3.74	3.98	4.16	4.31	4.44	4.55	4.65	4.73
	.01	3.76	4.28	4.59	4.82	4.99	5.13	5.25	5.36	5.45	5.53
120	.05	2.80	3.36	3.68	3.92	4.10	4.24	4.36	4.47	4.56	4.64
	.01	3.70	4.20	4.50	4.71	4.87	5.01	5.12	5.21	5.30	5.37
∞	.05	2.77	3.31	3.63	3.86	4.03	4.17	4.29	4.39	4.47	4.55
	.01	3.64	4.12	4.40	4.60	4.76	4.88	4.99	5.08	5.16	5.23

From: Table 11.2 in *The probability integrals of the range and of the Studentized range*, prepared by H. Leon Harter, Donald S. Clemm, and Eugene H. Guthrie. These tables are published in WADC Tech. Rep. 58–484, vol. 2, 1959, Wright Air Development Center, and are reproduced with the kind permission of the authors.

Note: Reject null hypothesis if obtained Q is equal to or greater than the tabled value.

TABLE 5. CRITICAL VALUES OF CHI SQUARE

df	α levels				
	.10	.05	.02	.01	.001
1	2.71	3.84	5.41	6.64	10.38
2	4.60	5.99	7.82	9.21	13.82
3	6.25	7.82	9.84	11.34	16.27
4	7.78	9.49	11.67	13.28	18.46
5	9.24	11.07	13.39	15.09	20.52
6	10.64	12.59	15.03	16.81	22.46
7	12.02	14.07	16.62	18.48	24.32
8	13.36	15.51	18.17	20.09	26.12
9	14.68	16.92	19.68	21.67	27.88
10	15.99	18.31	21.16	23.21	29.59
11	17.28	19.68	22.62	24.72	31.26
12	18.55	21.03	24.05	26.22	32.91
13	19.81	22.36	25.47	27.69	34.53
14	21.06	23.68	26.87	29.14	36.12
15	22.31	25.00	28.26	30.58	37.70
16	23.54	26.30	29.63	32.00	39.25
17	24.77	27.59	31.00	33.41	40.79
18	25.99	28.87	32.35	34.80	42.31
19	27.20	30.14	33.69	36.19	43.82
20	28.41	31.41	35.02	37.57	45.32
21	29.62	32.67	36.34	38.93	46.80
22	30.81	33.92	37.66	40.29	48.27
23	32.01	35.17	38.97	41.64	49.73
24	33.20	36.42	40.27	42.98	51.18
25	34.38	37.65	41.57	44.31	52.62
26	35.56	38.88	42.86	45.64	54.05
27	36.74	40.11	44.14	46.96	55.48
28	37.92	41.34	45.42	48.28	56.89
29	39.09	42.56	46.69	49.59	58.30
30	40.26	43.77	47.96	50.89	59.70

Source: Table F is taken from Table IV of Fisher and Yates, *Statistical Tables for Biological, Agricultural and Medical Research*, published by Longman Group Ltd., London (previously published by Oliver and Boyd, Ltd., Edinburgh), and by permission of the authors and publishers.

Note: Reject null hypothesis if obtained chi square is equal to or greater than the tabled value.

TABLE 6. CRITICAL VALUES FOR THE SPEARMAN RANK-ORDER CORRELATION COEFFICIENT

Number of Pairs, n	Level of Significance for a One-Tailed Test			
	.05	.025	.01	.005
	Level of Significance for a Two-Tailed Test			
	.10	.05	.02	.01
5	0.900	1.000	1.000	—
6	0.829	0.886	0.943	1.000
7	0.714	0.786	0.893	0.929
8	0.643	0.738	0.833	0.881
9	0.600	0.683	0.783	0.833
10	0.564	0.648	0.746	0.794
12	0.506	0.591	0.712	0.777
14	0.456	0.544	0.645	0.715
16	0.425	0.506	0.601	0.665
18	0.399	0.475	0.564	0.625
20	0.377	0.450	0.534	0.591
22	0.359	0.428	0.508	0.562
24	0.343	0.409	0.485	0.537
26	0.329	0.392	0.465	0.515
28	0.317	0.377	0.448	0.496
30	0.306	0.364	0.432	0.478

Source: Glasser, G. J., and R. F. Winter, "Critical Values of the Coefficient of Rank Correlation for Testing the Hypothesis of Independence." *Biometrika*, 48, 444 (1961).
Notes: Reject null hypothesis of obtained rho is equal to or greater than tabled value.

TABLE 7. CRITICAL VALUE OF *U* FOR THE MANN–WHITNEY *U* TEST

N of larger sample

N of smaller sample		p	2	3	4	5	6	7	8	9	10	11	12	13	14	15	16	17	18	19	20
	2	.001	0	0	0	0	0	0	0	0	0	0	0	0	0	0	0	0	0	0	0
		.005	0	0	0	0	0	0	0	0	0	0	0	0	0	0	0	0	0	1	1
		.01	0	0	0	0	0	0	0	0	0	0	0	1	1	1	1	1	1	2	2
		.025	0	0	0	0	0	0	1	1	1	1	2	2	2	2	2	3	3	3	3
		.05	0	0	0	1	1	1	2	2	2	2	3	3	4	4	4	4	5	5	5
		.10	0	1	1	2	2	2	3	3	4	4	5	5	5	6	6	7	7	8	8
	3	.001	0	0	0	0	0	0	0	0	0	0	0	0	0	0	1	1	1	1	1
		.005	0	0	0	0	0	0	1	1	1	2	2	2	3	3	3	3	4	4	4
		.01	0	0	0	0	1	1	2	2	2	3	3	3	4	4	5	5	5	5	6
		.025	0	0	1	2	2	3	3	4	4	5	5	6	6	7	7	8	8	8	9
		.05	0	1	1	2	3	3	4	5	5	6	6	7	8	8	9	10	10	11	12
		.10	1	2	2	3	4	5	6	6	7	8	9	10	11	11	12	13	14	15	16
	4	.001	0	0	0	0	0	0	0	1	1	1	2	2	2	3	3	4	4	4	4
		.005	0	0	0	1	1	2	2	3	3	4	4	5	6	6	7	7	8	9
		.01	0	0	1	2	2	3	4	4	5	6	6	7	9	8	9	10	10	11
		.025	0	0	1	2	3	4	5	5	6	7	8	9	10	11	12	12	13	14	15
		.05	0	1	2	3	4	5	6	7	8	9	10	11	12	13	15	16	17	18	19
		.10	1	2	4	5	6	7	8	10	11	12	13	14	16	17	18	19	21	22	23
	5	.001	0	0	0	0	0	0	1	2	2	3	3	4	4	5	6	6	7	8	8
		.005	0	0	0	1	2	2	3	4	5	6	7	8	8	9	10	11	12	13	14
		.01	0	0	1	2	3	4	5	6	7	8	9	10	11	12	13	14	15	16	17
		.025	0	1	2	3	4	6	7	8	9	10	12	13	14	15	16	18	19	20	21
		.05	1	2	3	5	6	7	9	10	12	13	14	16	17	19	20	21	23	24	26
		.10	2	3	5	6	8	9	11	13	14	16	18	19	21	23	24	26	28	29	31
	6	.001	0	0	0	0	0	2	3	4	5	5	6	7	8	9	10	11	12	13
		.005	0	0	1	2	3	4	5	6	7	8	10	11	12	13	14	16	17	18	19
		.01	0	0	2	3	4	5	7	8	9	10	12	13	14	16	17	19	20	21	23
		.025	0	2	3	4	6	7	9	11	12	14	15	17	18	20	22	23	25	26	28
		.05	1	3	4	6	8	9	11	13	15	17	18	20	22	24	26	27	29	31	33
		.10	2	4	6	8	10	12	14	16	18	20	22	24	26	28	30	32	35	37	39
	7	.001	0	0	0	1	2	3	4	6	7	8	9	10	11	12	14	15	16	17
		.005	0	0	1	2	4	5	7	8	10	11	13	14	16	17	19	20	22	23	25
		.01	0	1	2	4	5	7	8	10	12	13	15	17	18	20	22	24	25	27	29
		.025	0	2	4	6	7	9	11	13	15	17	19	21	23	25	27	29	31	33	35
		.05	1	3	5	7	9	12	14	16	18	20	22	25	27	29	31	34	36	38	40
		.10	2	5	7	9	12	14	17	19	22	24	27	29	32	34	37	39	42	44	47
	8	.001	0	0	0	1	2	3	5	6	7	9	10	12	13	15	16	18	19	21	22
		.005	0	0	2	3	5	7	8	10	12	14	16	18	19	21	23	25	27	29	31
		.01	0	1	3	5	7	8	10	12	14	16	18	21	23	25	27	29	31	33	35
		.025	1	3	5	7	9	11	14	16	18	20	23	25	27	30	32	35	37	39	42
		.05	2	4	6	9	11	14	16	19	21	24	27	29	32	34	37	40	42	45	48
		.10	3	6	8	11	14	17	20	23	25	28	31	34	37	40	43	46	49	52	55
	9	.001	0	0	0	2	3	4	6	8	9	11	13	15	16	18	20	22	24	26	27
		.005	0	1	2	4	6	8	10	12	14	17	19	21	23	25	28	30	32	34	37
		.01	0	2	4	6	8	10	12	15	17	19	22	24	27	29	32	34	37	39	41
		.025	1	3	5	8	11	13	16	18	21	24	27	29	32	35	38	40	43	46	49
		.05	2	5	7	10	13	16	19	22	25	28	31	34	37	40	43	46	49	52	55
		.10	3	6	10	13	16	19	23	26	29	32	36	39	42	46	49	53	56	59	63

Source: Abridged from L. R. Verdooren, "Extended Tables of Critical Values for Wilcoxon's Test Statistic," *Biometrika* 50 (1963), pp. 177–86.

Note: Reject null hypothesis if obtained *U* is less than the tabled value.

TABLE 7. (CONTINUED)

N of larger sample

	p	2	3	4	5	6	7	8	9	10	11	12	13	14	15	16	17	18	19	20
10	.001	0	0	1	2	4	6	7	9	11	13	15	18	20	22	24	26	28	30	33
	.005	0	1	3	5	7	10	12	14	17	19	22	25	27	30	32	35	38	40	43
	.01	0	2	4	7	9	12	14	17	20	23	25	28	31	34	37	39	42	45	48
	.025	1	4	6	9	12	15	18	21	24	27	30	34	37	40	43	46	49	53	56
	.05	2	5	8	12	15	18	21	25	28	32	35	38	42	45	49	52	56	59	63
	.10	4	7	11	14	18	22	25	29	33	37	40	44	48	52	55	59	63	67	71
11	.001	0	0	1	3	5	7	9	11	13	16	18	21	23	25	28	30	33	35	38
	.005	0	1	3	6	8	11	14	17	19	22	25	28	31	34	37	40	43	46	49
	.01	0	2	5	8	10	13	16	19	23	26	29	32	35	38	42	45	48	51	54
	.025	1	4	7	10	14	17	20	24	27	31	34	38	41	45	48	52	56	59	63
	.05	2	6	9	13	17	20	24	28	32	35	39	43	47	51	55	58	62	66	70
	.10	4	8	12	16	20	24	28	32	37	41	45	49	53	58	62	66	70	74	79
12	.001	0	0	1	3	5	8	10	13	15	18	21	24	26	29	32	35	38	41	43
	.005	0	2	4	7	10	13	16	19	22	25	28	32	35	38	42	45	48	52	55
	.01	0	3	6	9	12	15	18	22	25	29	32	36	39	43	47	50	54	57	61
	.025	2	5	8	12	15	19	23	27	30	34	38	42	46	50	54	58	62	66	70
	.05	3	6	10	14	18	22	27	31	35	39	43	48	52	56	61	65	69	73	78
	.10	5	9	13	18	22	27	31	36	40	45	50	54	59	64	68	73	78	82	87
13	.001	0	0	2	4	6	9	12	15	18	21	24	27	30	33	36	39	43	46	49
	.005	0	2	4	8	11	14	18	21	25	28	32	35	39	43	46	50	54	58	61
	.01	1	3	6	10	13	17	21	24	28	32	36	40	44	48	52	56	60	64	68
	.025	2	5	9	13	17	21	25	29	34	38	42	46	51	55	60	64	68	73	77
	.05	3	7	11	16	20	25	29	34	38	43	48	52	57	62	66	71	76	81	85
	.10	5	10	14	19	24	29	34	39	44	49	54	59	64	69	75	80	85	90	95
14	.001	0	0	2	4	7	10	13	16	20	23	26	30	33	37	40	44	47	51	55
	.005	0	2	5	8	12	16	19	23	27	31	35	39	43	47	51	55	59	64	68
	.01	1	3	7	11	14	18	23	27	31	35	39	44	48	52	57	61	66	70	74
	.025	2	6	10	14	18	23	27	32	37	41	46	51	56	60	65	70	75	79	84
	.05	4	8	12	17	22	27	32	37	42	47	52	57	62	67	72	78	83	88	93
	.10	5	11	16	21	26	32	37	42	48	53	59	64	70	75	81	86	92	98	103
15	.001	0	0	2	5	8	11	15	18	22	25	29	33	37	41	44	48	52	56	60
	.005	0	3	6	9	13	17	21	25	30	34	38	43	47	52	56	61	65	70	74
	.01	1	4	8	12	16	20	25	29	34	38	43	48	52	57	62	67	71	76	81
	.025	2	6	11	15	20	25	30	35	40	45	50	55	60	65	71	76	81	86	91
	.05	4	8	13	19	24	29	34	40	45	51	56	62	67	73	78	84	89	95	101
	.10	6	11	17	23	28	34	40	46	52	58	64	69	75	81	87	93	99	105	111
16	.001	0	0	3	6	9	12	16	20	24	28	32	36	40	44	49	53	57	61	66
	.005	0	3	6	10	14	19	23	28	32	37	42	46	51	56	61	66	71	75	80
	.01	1	4	8	13	17	22	27	32	37	42	47	52	57	62	67	72	77	83	88
	.025	2	7	12	16	22	27	32	38	43	48	54	60	65	71	76	82	87	93	99
	.05	4	9	15	20	26	31	37	43	49	55	61	66	72	78	84	90	96	102	108
	.10	6	12	18	24	30	37	43	49	55	62	68	75	81	87	94	100	107	113	120
17	.001	0	1	3	6	10	14	18	22	26	30	35	39	44	48	53	58	62	67	71
	.005	0	3	7	11	16	20	25	30	35	40	45	50	55	61	66	71	76	82	87
	.01	1	5	9	14	19	24	29	34	39	45	50	56	61	67	72	78	83	89	94
	.025	3	7	12	18	23	29	35	40	46	52	58	64	70	76	82	88	94	100	106
	.05	4	10	16	21	27	34	40	46	52	58	65	71	78	84	90	97	103	110	116
	.10	7	13	19	26	32	39	46	53	59	66	73	80	86	93	100	107	114	121	128

N of smaller sample

TABLE 7. (CONTINUED)

N of larger sample

N of smaller sample	p	2	3	4	5	6	7	8	9	10	11	12	13	14	15	16	17	18	19	20
18	.001	0	1	4	7	11	15	19	24	28	33	38	43	47	52	57	62	67	72	77
	.005	0	3	7	12	17	22	27	32	38	43	48	54	59	65	71	76	82	88	93
	.01	1	5	10	15	20	25	31	37	42	48	54	60	66	71	77	83	89	95	101
	.025	3	8	13	19	25	31	37	43	49	56	62	68	75	81	87	94	100	107	113
	.05	5	10	17	23	29	36	42	49	56	62	69	76	83	89	96	103	110	117	124
	.10	7	14	21	28	35	42	49	56	63	70	78	85	92	99	107	114	121	129	136
19	.001	0	1	4	8	12	16	21	26	30	35	41	46	51	56	61	67	72	78	83
	.005	1	4	8	13	18	23	29	34	40	46	52	58	64	70	75	82	88	94	100
	.01	2	5	10	16	21	27	33	39	45	51	57	64	70	76	83	89	95	102	108
	.025	3	8	14	20	26	33	39	46	53	59	66	73	79	86	93	100	107	114	120
	.05	5	11	18	24	31	38	45	52	59	66	73	81	88	95	102	110	117	124	131
	.10	8	15	22	29	37	44	52	59	67	74	82	90	98	105	113	121	129	136	144
20	.001	0	1	4	8	13	17	22	27	33	38	43	49	55	60	66	71	77	83	89
	.005	1	4	9	14	19	25	31	37	43	49	55	61	68	74	80	87	93	100	106
	.01	2	6	11	17	23	29	35	41	48	54	61	68	74	81	88	94	101	108	115
	.025	3	9	15	21	28	35	42	49	56	63	70	77	84	91	99	106	113	120	128
	.05	5	12	19	26	33	40	48	55	63	70	78	85	93	101	108	116	124	131	139
	.10	8	16	23	31	39	47	55	63	71	79	87	95	103	111	120	128	136	144	152

TABLE 8. CRITICAL VALUES OF W FOR THE WILCOXON TEST

	Level of Significance for a One-Tailed Test					Level of Significance for a One-Tailed Test			
	0.05	0.025	0.01	0.005		0.05	0.025	0.01	0.005
	Level of Significance for a Two-Tailed Test					Level of Significance for a Two-Tailed Test			
N	0.10	0.05	0.02	0.01	N	0.10	0.05	0.02	0.01
5	0	—	—	—	28	130	116	101	91
6	2	0	—	—	29	140	126	110	100
7	3	2	0	—	30	151	137	120	109
8	5	3	1	0	31	163	147	130	118
9	8	5	3	1	32	175	159	140	128
10	10	8	5	3	33	187	170	151	138
11	13	10	7	5	34	200	182	162	148
12	17	13	9	7	35	213	195	173	159
13	21	17	12	9	36	227	208	185	171
14	25	21	15	12	37	241	221	198	182
15	30	25	19	15	38	256	235	211	194
16	35	29	23	19	39	271	249	224	207
17	41	34	27	23	40	286	264	238	220
18	47	40	32	27	41	302	279	252	233
19	53	46	37	32	42	319	294	266	247
20	60	52	43	37	43	336	310	281	261
21	67	58	49	42	44	353	327	296	276
22	75	65	55	48	45	371	343	312	291
23	83	73	62	54	46	389	361	328	307
24	91	81	69	61	47	407	378	345	322
25	100	89	76	68	48	426	396	362	339
26	110	98	84	75	49	446	415	379	355
27	119	107	92	83	50	466	434	397	373

Source: From F. Wilcoxon, S. Katte, and R. A. Wilcox, *Critical Values and Probability Levels for the Wilcoxon Rank Sum Test and the Wilcoxon Signed Rank Test*, New York, American Cyanamid Co., 1963, and F. Wilcoxon and R. A. Wilcox, *Some Rapid Approximate Statistical Procedures*, New York, Lederle Laboratories, 1964 as used in Runyon and Haber, *Fundamentals of Behavioral Statistics*, 1967, Addison-Wesley, Reading, Mass.

Note: Reject null hypothesis if obtained W is equal to or less than tabled value.

TABLE 9. TABLE OF PROBABILITIES ASSOCIATED WITH VALUES AS LARGE AS OBSERVED VALUES OF χ_r^2 IN THE FRIEDMAN TWO-WAY ANALYSIS OF VARIANCE BY RANKS*

TABLE N_1. $k = 3$

$N = 2$		$N = 3$		$N = 4$		$N = 5$	
χ_r^2	p	χ_r^2	p	χ_r^2	p	χ_r^2	p
0	1.000	.000	1.000	.0	1.000	.0	1.000
1	.833	.667	.944	.5	.931	.4	.954
3	.500	2.000	.528	1.5	.653	1.2	.691
4	.167	2.667	.361	2.0	.431	1.6	.522
		4.667	.194	3.5	.273	2.8	.367
		6.000	.028	4.5	.125	3.6	.182
				6.0	.069	4.8	.124
				6.5	.042	5.2	.093
				8.0	.0046	6.4	.039
						7.6	.024
						8.4	.0085
						10.0	.00077

$N = 6$		$N = 7$		$N = 8$		$N = 9$	
χ_r^2	p	χ_r^2	p	χ_r^2	p	χ_r^2	p
.00	1.000	.000	1.000	.00	1.000	.000	1.000
.33	.956	.286	.964	.25	.967	.222	.971
1.00	.740	.857	.768	.75	.794	.667	.814
1.33	.570	1.143	.620	1.00	.654	.889	.865
2.33	.430	2.000	.486	1.75	.531	1.556	.569
3.00	.252	2.571	.305	2.25	.355	2.000	.398
4.00	.184	3.429	.237	3.00	.285	2.667	.328
4.33	.142	3.714	.192	3.25	.236	2.889	.278
5.33	.072	4.571	.112	4.00	.149	3.556	.187
6.33	.052	5.429	.085	4.75	.120	4.222	.154
7.00	.029	6.000	.052	5.25	.079	4.667	.107
8.33	.012	7.143	.027	6.25	.047	5.556	.069
9.00	.0081	7.714	.021	6.75	.038	6.000	.057
9.33	.0055	8.000	.016	7.00	.030	6.222	.048
10.33	.0017	8.857	.0084	7.75	.018	6.889	.031
12.00	.00013	10.286	.0036	9.00	.0099	8.000	.019
		10.571	.0027	9.25	.0080	8.222	.016
		11.143	.0012	9.75	.0048	8.667	.010
		12.286	.00032	10.75	.0024	9.556	.0060
		14.000	.000021	12.00	.0011	10.667	.0035
				12.25	.00086	10.889	.0029
				13.00	.00026	11.556	.0013
				14.25	.000061	12.667	.00066
				16.00	.0000036	13.556	.00035
						14.000	.00020
						14.222	.000097
						14.889	.000054
						16.222	.000011
						18.000	.0000006

* Adapted from Friedman, M. 1937. The use of ranks to avoid the assumption of normality implicit in the analysis of variance. *J. Amer. Statist. Ass.*, **32**, 688–689, with the kind permission of the author and the publisher.

TABLE 9 (CONT'D). TABLE OF PROBABILITIES ASSOCIATED WITH VALUES AS LARGE AS OBSERVED VALUES OF χ_r^2 IN THE FRIEDMAN TWO-WAY ANALYSIS OF VARIANCE BY RANKS* (CONTINUED)

TABLE N_{II}. $k = 4$

$N = 2$		$N = 3$		$N = 4$			
χ_r^2	p	χ_r^2	p	χ_r^2	p	χ_r^2	p
.0	1.000	.2	1.000	.0	1.000	5.7	.141
.6	.958	.6	.958	.3	.992	6.0	.105
1.2	.834	1.0	.910	.6	.928	6.3	.094
1.8	.792	1.8	.727	.9	.900	6.6	.077
2.4	.625	2.2	.608	1.2	.800	6.9	.068
3.0	.542	2.6	.524	1.5	.754	7.2	.054
3.6	.458	3.4	.446	1.8	.677	7.5	.052
4.2	.375	3.8	.342	2.1	.649	7.8	.036
4.8	.208	4.2	.300	2.4	.524	8.1	.033
5.4	.167	5.0	.207	2.7	.508	8.4	.019
6.0	.042	5.4	.175	3.0	.432	8.7	.014
		5.8	.148	3.3	.389	9.3	.012
		6.6	.075	3.6	.355	9.6	.0069
		7.0	.054	3.9	.324	9.9	.0062
		7.4	.033	4.5	.242	10.2	.0027
		8.2	.017	4.8	.200	10.8	.0016
		9.0	.0017	5.1	.190	11.1	.00094
				5.4	.158	12.0	.000072

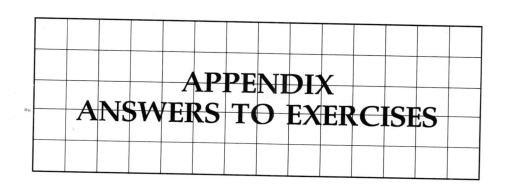

APPENDIX
ANSWERS TO EXERCISES

CHAPTER 2

1.
Flavor	f
Vanilla	9
Chocolate	7
Strawberry	5
Rocky Road	4

2. Nominal level.

3.
Flavor	Rank
Vanilla	1
Chocolate	2
Strawberry	3
Rocky Road	4

4. Ordinal level.

5. X = raw score, f = frequency, N = number of scores, ll = real lower limit.

6.
Score	f	Score	f	Score	f
28	1	21	2	14	2
27	1	20	2	13	1
26	1	19	3	12	0
25	0	18	1	11	0
24	1	17	1	10	0
23	1	16	1	9	1
22	2	15	2		

7.
Interval	f
27–29	2
24–26	2
21–23	5
18–20	6
15–17	4
12–14	3
9–11	1

8. Interval level.

9.
Interval	f
41–44	2
37–40	4
33–36	6
29–32	3
25–28	0
21–24	2
	$N = 17$

10. 11.5, 19.5, 16.5, 99.5

CHAPTER 3

1. | X | f |
    |----|---|
    | 30 | 1 |
    | 29 | 2 |
    | 28 | 6 |
    | 27 | 5 |
    | 26 | 2 |
    | 25 | 2 |
    | 24 | 0 |
    | 23 | 4 |

    Mode = 28, Median = 27, Mean = 26.591

2. | X | f |
    |----|---|
    | 63 | 2 |
    | 62 | 4 |
    | 61 | 6 |
    | 60 | 5 |
    | 59 | 3 |
    | 58 | 3 |
    | 57 | 1 |

    Mode = 61, Median = 60.5, Mean = 60.333

3. | Number of trials | f |
    |------------------|---|
    | 6 | 3 |
    | 5 | 8 |
    | 4 | 6 |
    | 3 | 4 |
    | 2 | 2 |
    | 1 | 1 |

    Mode = 5, Median = 4, Mean = 4.125

4. Mode = 119 Median = 115.5 Mean = 114.859

Interval	f
125–129	6
120–124	10
115–119	29
110–114	15
105–109	11
100–104	7

CHAPTER 4

1.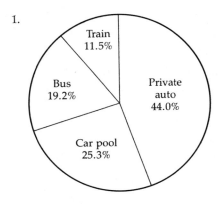

    Segment of circle:
    Private auto   158.4°
    Car pool   91.1°
    Bus   69.1°
    Train   41.4°

221
APPENDIX ANSWERS TO EXERCISES

(or)

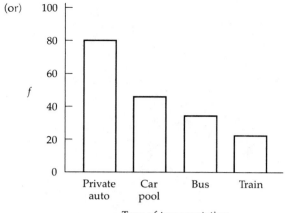

2.

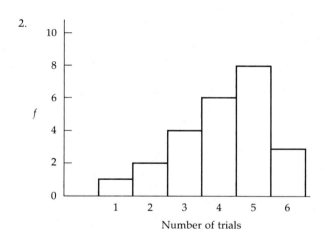

(or)

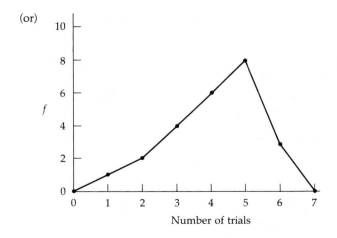

222
APPENDIX
ANSWERS TO
EXERCISES

3.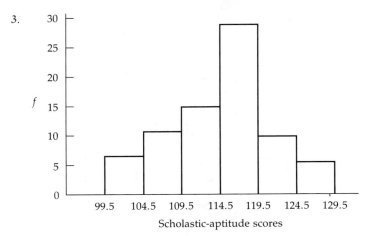

(or)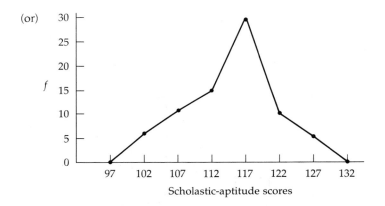

CHAPTER 5
1. Mean of a population 2. Statistics
3. Parameters 4. Roman 5. Greek
6. A sample in which every member of a population has an equal chance of being selected for the sample.

CHAPTER 6
1. s is estimate of population standard deviation
s^2 is estimate of population variance
x is the deviation of a score from the mean
σ is the population standard deviation
σ^2 is the population variance

2. a) range = 6   b) average deviation = 1.385   c) and d) variance = 2.960
   e) standard deviation = 1.720   f) mean = 51.000   g) 2.000   h) −2.000

3. a) range = 13   b) average deviation = 3.250   c) and d) variance = 14.110
   e) standard deviation = 3.756   f) mean = 1.464   g) 4.536   h) −2.464

4. a) range = 17   b) variance = 16.575   c) standard deviation = 4.070
   d) mean = 18.667   e)

Interval	f
25–27	3
22–24	11
19–21	20
16–18	14
13–15	6
10–12	6

5. a) range = 35   b) variance = 73.492   c) standard deviation = 8.573
   d) mean = 70.612   e)   f) 12.388   g) −15.612

Interval	f
86–90	3
81–85	4
76–80	6
71–75	12
66–70	10
61–65	8
56–60	4
51–55	2

CHAPTER 7

1. a) 4   b) $z = -1$ c) .0668   d) $z = 1.75$ e) .0401   f) .5859
   g) .1747   h) .1498

2. a) 20   b) $z = .05$ c) .1469   d) $z = .05$ e) .4801   f) .8356
   g) .3811   h) .0455

CHAPTER 8

1. probability
2. a probability distribution
3. $P = .4332$
4. $P = .4878$
5. $P = .7888$
6. $P = .0401$
7. $P = .0228$

CHAPTER 9

1. $\sigma_{\bar{x}}$ represents the population standard error of the mean, determined by calculating the standard deviation of all possible means of the same sized samples from a population.
   $s_{\bar{x}}$ represents an estimate of the population standard error of the mean, determined from the data in one sample from that population.

2. a) 1.778   b) .0122   c) .0455   d) .0106   e) .1299   f) .7372
3. a) 1.500   b) .9994   c) .5000   d) .3069   e) .3108   f) .1138

CHAPTER 10

1. 57.744 − 61.322
2. 57.094 − 61.973
3. 17.280 − 18.720
4. 17.001 − 18.999
5. 17.595 − 19.739
6. 17.222 − 20.112
7. 73.727 − 76.273
8. 73.314 − 76.686

CHAPTER 11

1. a) $r = .903$
   b) Coefficient of Determination $= .816$
   This indicates that about 82% of the variance in the mechanical comprehension scores is associated with the variance in the divergent thinking scores.

   c)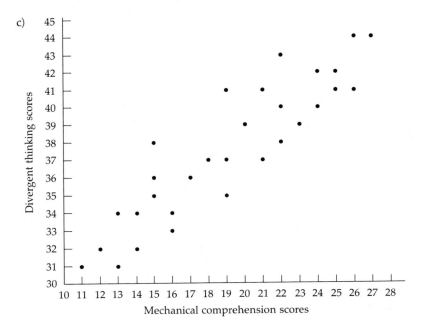

2. a) $r = -.473$
   b) Coefficient of Determination $= .223$
   This indicates that about 22% of the variance in the spelling scores is associated with the variance in the times of taking the test.

   c)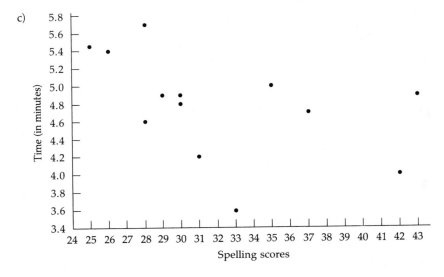

CHAPTER 12

1. a) $\tilde{Y} = 8.89 + .571X$ b) For $X = 15$: $\tilde{Y} = 17.458$
   For $X = 40$: $\tilde{Y} = 31.738$

c)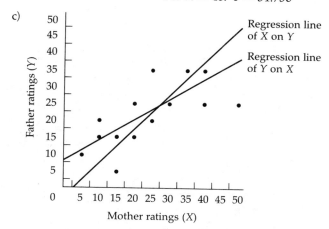

d) $\tilde{X} = 4.037 + .936Y$ e) For $Y = 20$: $\tilde{X} = 22.76$
   For $Y = 45$: $\tilde{X} = 46.162$
f) Regression line of X on Y shown in answer to Exercise 1c.

2. a) $\tilde{Y} = -1.278 + .806X_1 + .067X_2$ b) $\tilde{Y} = 2.197$ c) $\tilde{Y} = 3.405$
3. a) $\tilde{Y} = 23.228 + .749X$ b) For $X = 14$: $\tilde{Y} = 33.720$
   For $X = 24$: $\tilde{Y} = 41.214$
   c) $\tilde{X} = -21.784 + 1.089Y$ d) For $Y = 32$: $\tilde{X} = 13.070$
   For $Y = 42$: $\tilde{X} = 23.962$
   e) and f)

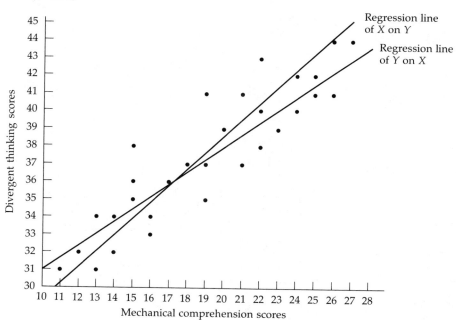

225

4. a) $\tilde{Y} = 59.312 + 1.663(X_1) - .81(X_2)$
   b) For $X_1 = 15$ and $X_2 = 27$: $\tilde{Y} = 62.387$
      For $X_1 = 22$ and $X_2 = 35$: $\tilde{Y} = 67.548$

CHAPTER 13

1. There is a difference in effectiveness between the variable method and the constant method for teaching chickens to peck at a red circle.
2. There is no difference in effectiveness between the variable method and the constant method for teaching chickens to peck at a red circle.
3. $\bar{X}_1 - \bar{X}_2 = -10$; cutoff point: $-1.96 \times 4 = -7.84$; reject the null hypothesis.
4. $\bar{X}_1 - \bar{X}_2 = 6$; cutoff point: $1.96 \times 4 = 7.84$; do not reject the null hypothesis.
5. There is a difference in the levels of scores students obtain on group A exercises and group B exercises.
6. There is no difference in the levels of scores students obtain on group A exercises and group B exercises.
7. $\bar{X}_1 - \bar{X}_2 = 4$; cutoff point: $1.96 \times 2.5 = 4.9$; do not reject the null hypothesis.
8. $\bar{X}_1 - \bar{X}_2 = -7$; cutoff point: $-1.96 \times 2.5 = -4.9$; reject the null hypothesis.

CHAPTER 14

1. t test for independent samples—variances assumed equal
   Intensive $\bar{X} = 39.524$ Periodic $\bar{X} = 48.476$ $t = -9.914$ $df = 40$
   Table 2 gives critical $t = 2.021$
   Computer calculates $P < .001$
   Reject null hypothesis.

2. t test for dependent samples
   Form A $\bar{X} = 100.050$ Form B $\bar{X} = 99.900$
   $t = .258$ $df = 19$
   Table 2 gives critical $t = 2.861$
   Computer calculates $P = .999$
   Do not reject null hypothesis.

3. t test for independent samples—variances assumed unequal
   Drug XX-7 $\bar{X} = 79.833$ Drug XX-8 $\bar{X} = 93.222$
   $t = -8.495$ $df = 10$ (Welch's technique)
   Table 2 gives critical $t = 2.228$
   Computer calculates $P < .001$
   Reject null hypothesis.

CHAPTER 15

1. Do not reject the null hypothesis.
2. Probability of a Type II error: $\beta = .4602$; power: $1 - \beta = .5398$.
3. Probability of a Type II error: $\beta = .7794$; power: $1 - \beta = .2206$.
4. Do not reject the null hypothesis.
5. Probability of a Type II error: $\beta = .0154$; power: $1 - \beta = .9846$.
6. Probability of a Type II error: $\beta = .9789$; power: $1 - \beta = .0211$.
7. $r = .903$ $df = 28$ $t = 11.153$ $P < .001$
   Reject the null hypothesis.

CHAPTER 16

1. One-Way Analysis of Variance—Independent Samples

	Sum of squares	df	Mean square	F
Between groups	398.506	2	199.253	42.664
Within groups	84.065	18	4.670	
Total	482.571	20		

From Table 3, critical $F = 3.55$
Reject null hypothesis.

Group Means: Individual Instruction $\bar{X} = 26.625$
 Small Group Instruction $\bar{X} = 30.143$
 Large Group Instruction $\bar{X} = 37.333$

Scheffé's test for multiple comparisons is considered significant if the F ratio is as large as or larger than 7.1.

Indiv. vs. small group	$F = 9.893$	significant
Indiv. vs. large group	$F = 84.181$	significant
Small group vs. large group	$F = 35.766$	significant

Conclusion: Each method is significantly different from the others.

2. Using the Pearson correlation, $r = .903$, $t = 11.153$, $P < .001$
Reject the null hypothesis.
Conclusion: There is a significant relationship between the two variables.

3. One-Way Analysis of Variance—Independent Samples

	Sum of squares	df	Mean square	F
Between groups	584.458	3	194.819	38.770
Within groups	100.500	20	5.025	
Total	684.958	23		

From Table 3, critical $F = 4.94$
Reject null hypothesis.

Group Means: Diet 1 $\bar{X} = 12.500$
 Diet 2 $\bar{X} = 12.167$
 Diet 3 $\bar{X} = 23.833$
 Diet 4 $\bar{X} = 12.667$

Tukey's test for multiple samples is considered significant if Q is as large as or larger than 5.02.

Diet 1 vs. diet 2	$Q = .364$	
Diet 1 vs. diet 3	$Q = 12.384$	significant
Diet 1 vs. diet 4	$Q = .182$	
Diet 2 vs. diet 3	$Q = 12.748$	significant
Diet 2 vs. diet 4	$Q = .546$	
Diet 3 vs. diet 4	$Q = 12.202$	significant

Conclusion: Diet 3 has a significantly higher mean than all of the other three diets. Diets 1, 2, and 4 do not differ among themselves.

4. One-Way Analysis of Variance—Independent Samples

	Sum of squares	df	Mean square	F
Between groups	701.618	2	350.809	57.758
Within groups	546.640	90	6.074	
Total	1248.258	92		

From Table 3, critical $F = 3.11$
Reject null hypothesis.
Group Means: Lecture only $\bar{X} = 32.135$
              Lecture-part. $\bar{X} = 38.600$
              Participation only $\bar{X} = 35.731$

Scheffé's test for multiple comparisons is considered significant if the F ratio is as large as or larger than 6.22.

Lecture only vs. lecture-part.	$F = 114.001$	significant
Lecture only vs. participation only	$F = 32.503$	significant
Lecture-part. vs. participation only	$F = 18.879$	significant

Conclusion: Each instructional style is significantly different from the others.

5. One-Way Analysis of Variance—Independent Samples

	Sum of squares	df	Mean square	F
Between groups	750.700	3	250.233	39.265
Within groups	739.266	116	6.373	
Total	1489.967	119		

From Table 3, critical $F = 3.98$
Reject null hypothesis.
Group Means: Brand 1 $\bar{X} = 50.733$
              Brand 2 $\bar{X} = 51.300$
              Brand 3 $\bar{X} = 56.200$
              Brand 4 $\bar{X} = 49.700$

Tukey's test for multiple samples is considered significant if Q is as large as or larger than 4.60.

Brand 1 vs. brand 2	$Q = 1.229$	
Brand 1 vs. brand 3	$Q = 11.861$	significant
Brand 1 vs. brand 4	$Q = 2.242$	
Brand 2 vs. brand 3	$Q = 10.631$	significant
Brand 2 vs. brand 4	$Q = 3.471$	
Brand 3 vs. brand 4	$Q = 14.103$	significant

Conclusion: Brand 3 is significantly superior to the other three brands, which do not differ among themselves.

CHAPTER 18

1. Goodness-of-fit test. $\chi^2 = 3.01$, $df = 1$
   Table 5: Critical chi square = 6.64
   Do not reject null hypothesis.

2. Test of Independence. $\chi^2 = 4.727$, $df = 1$
   Table 5: Critical chi square = 3.84
   Reject null hypothesis.

3. Test of Independence. $\chi^2 = 9.095$, $df = 6$
   Table 5: Critical chi square = 16.81
   Do not reject null hypothesis.

4. Test of Independence. $\chi^2 = 3.763$, $df = 1$
   Table 5: Critical chi square = 3.84
   Do not reject null hypothesis.

5. Goodness-of-fit test. $\chi^2 = 22.707$, $df = 2$
   Table 5: Critical chi square = 5.99
   Reject null hypothesis.

6. Test of Independence. $\chi^2 = 12.876$, $df = 2$
   Table 5: Critical chi square = 4.60
   Reject null hypothesis.

CHAPTER 19

1. Kruskal–Wallis H test.
   $H = 13.530$ (corrected for ties)
   For chi square = 13.530, $df = 3$
   Table 5 gives critical chi square = 11.34
   Reject null hypothesis.

2. Spearman rank correlation.
   rho = .922 (corrected for ties)
   $N = 11$
   Table 6 gives critical rho = .537 (directional test)
   Reject null hypothesis.

3. Wilcoxon matched-pairs signed-ranks test.
   $T = 44$ $N = 14$
   Table 8 gives critical $T = 12$
   Do not reject null hypothesis.

4. Friedman two-way analysis of variance by ranks for repeated measures.
   chi square = 10.4, $df = 2$
   Table 5 gives critical chi square = 5.99
   Reject null hypothesis.

5. Mann–Whitney U test.
   $U = 3.5$
   Table 7 gives critical $U = 14$
   Reject null hypothesis.

6. Spearman rank correlation.
   rho = $-.315$ (corrected for ties)
   $N = 9$
   Table 6 gives critical rho = .700
   Do not reject null hypothesis.

7. Kruskal–Wallis H test.
   $H = 2.615$ (corrected for ties)
   For chi square $= 2.615$, $df = 2$
   Table 5 gives critical chi square $= 5.99$
   Do not reject null hypothesis.

8. Mann–Whitney U test.
   $U = 29$
   Table 7 gives critical $U = 25$
   Do not reject null hypothesis.

GLOSSARY OF SYMBOLS AND ABBREVIATIONS

a	Intercept in the regression equation
A.D.	Average deviation
b_{xy}	Regression coefficient for variable X on variable Y
b_{yx}	Regression coefficient for variable Y on variable X
D	Difference between two scores or ranks
df	Degrees of freedom
E	Expected frequency (in chi-square test)
F	Test statistic for the ratio of two variances
f	Frequency
H	Test statistic for Kruskal-Wallis ANOVA by ranks test
$\ell\ell$	Lower limits of a class interval
MS	Mean square (in analysis of variance)
N	Number of scores or ranks
O	Observed frequency (in chi-square test)
P	Probability
Q	Test statistic for Tukey method for multiple comparison
r	Pearson product-moment correlation coefficient
r^2	Coefficient of determination
$s_{\bar{D}}$	Estimate of population standard error of mean difference
s	Estimate of population standard deviation
s^2	Estimate of population variance
$s_{\bar{X}}$	Estimate of the standard error of the mean
s_D^2	Estimate of the population variance of difference scores
$s_{\bar{X}_1 - \bar{X}_2}$	Estimate of the standard error of the difference between means
SS_b	Sum of squares between groups (in analysis of variance)
SS_w	Sum of squares within groups (in analysis of variance)
SS_t	Sum of squares of total groups (in analysis of variance)
t	Student's t-test statistic
U	Test statistic for Mann-Whitney U test
X	Score value
$\bar{X}$	Sample mean
$\tilde{X}$	Estimate value (in regression equation)

x	Deviation of a score from the mean
$\lvert x \rvert$	Absolute deviation score
$\tilde{Y}$	Estimated value (in regression equation)
z	Relative deviate
α	Probability of making a Type I error
β	Probability of making a Type II error
Σ	The sum of
μ	Population mean
σ	Population standard deviation
$\sigma_{\bar{x}}$	Standard error of the mean
σ^2	Population variance
$\sigma_{\bar{x}_1 - \bar{x}_2}$	Standard error of the difference between means
χ^2	Chi-square test statistic
ρ	Rank-order correlation coefficient

INDEX

Absolute deviation score, 40
Analysis of covariance, 162
Analysis of variance
   Friedman test, 194
   one-way, 141
   repeated measures, 166
   two-way, 157
Average deviation, 40

Bar graph, 29
Best-fit line, 41
Bimodal curve, 32

Central Limit Theorem, 63
Central tendency, 17
Chi square, 169
   test for equality of proportions, 177
   test for goodness of fit, 170
   test for independence of two variables, 175
Class interval, 14
Coefficient of correlation
   Pearson product-moment correlation, 84, 86
   rank-order correlation, 184
Coefficient of determination, 88
Confidence interval, 69
   ninety-five percent, 71
   ninety-nine percent, 73
Correlation, 81
   negative, 83
   perfect, 83
   positive, 83
   rank-order, 184
   zero, 85
Covariance analysis, 162

Data, 5
Degrees of freedom, 44

Descriptive statistics, 2
Deviation score, 40
Distribution, normal, 49
   probability, 58
   sample means, 63

Expected frequencies in chi-square test, 172

F distribution, 147
F ratio
   analysis of variance, 142
   difference between two variance estimates, 147
Fisher, R. A., 122
Frequency distribution, 11
   for grouped data, 14
   for ungrouped data, 13
Frequency of scores, 11
Frequency polygon, 31
Friedman two-way ANOVA by ranks, 194

Goodness-of-fit test, 170
Gossett, W. S., 75

Histogram, bar, 29
Hypothesis, 13
   alternate, 128
   directional, 129, 133
   nondirectional, 129, 133
   null, 106
   research, 106
   testing, 105

Independent means, 113
Inferential statistics, 2, 105
Interval estimate, 71
Interval measurement, 9

Kruskal-Wallis one-way ANOVA by ranks, 193

Limits, real, 13
Lower limit of class interval, 14

Mann-Whitney U test, 189
Mean, 20
Mean square, 142, 146
Measurement
  continuous, 10
  discrete, 10
  interval, 9
  nominal, 6
  ordinal, 8, 183
  ratio, 9
Median, 18
Mode, 17
Multiple comparisons, 150
Multiple regression, 99
  equation, 100

Nominal measurement, 6
Nonindependent means, 95
Nonparametric tests, 169
Normal curve, 49
Normal distribution, 49
Null hypothesis, 106, 127

Observed frequencies in chi-square test, 172
One-tail test, 128
Ordinal measurement, 5, 183

Parameter, 37
Parametric tests, 169
Partial regression coefficient, 100
Pearson, K., 84
Pearson product-moment correlation, 84, 86
Pie chart, 27
Point estimate, 71
Polygon, frequency, 31
Population, 3
Power of a statistical test, 136
Probability, 57

Range, 39
Rank order, 12
  coefficient, 185
  correlation, 184
Ratio measurement, 9
Real limits of a class interval, 14
Regression, 91
  coefficient, 93
  equation, 91, 93
  line, 93
  multiple, 99
Relative deviate, 53
Relative frequency, 57
Rho, 184

Sample, 36
  random, 36
Sampling distribution of means, 64
Sampling error, 65
Scales of measurement, 5
Scatter diagram, 82
Scheffé method for comparisons, 152
Significance level, 134
Skewed distribution
  negative, 32
  positive, 32
Smooth curve, 32
Spearman rank correlation, 184
Standard deviation, 45, 51
Standard error of the difference between means, 109, 115
Standard error of the mean, 64, 66
Statistic, 37
Student's distributions of t, 75
Sum of squares, 42, 145, 159

t distribution, 75
t test
  for independent means, 114, 119
  for nonindependent means, 121
Tukey method for multiple comparisons, 151
Two-tail test, 128
Type I and Type II errors, 135

Upper limit of a class interval, 14

Variability, 39
Variable, 5
  continuous, 10
  dependent, 92
  discrete, 10
  independent, 92
Variance, 41
  between-groups estimate, 142
  estimate of population, 43
  within-groups estimate, 142

Welch, B. L., 119
Wilcoxon matched-pairs signed-ranks test, 191

Yates' correction for continuity, 174